Praise for *T-Minus AI*

"Mike Kanaan is an influential new voice in the field of AI, and his thoughts paint an insightful perspective. A thought-provoking read."
— ERIC SCHMIDT, former CEO and executive chairman of Google

"Too many discussions of artificial intelligence are dominated by idealists and cynics. Mike Kanaan is neither: He's a realist with a wealth of insight on how smart machines are shaping the future. This is one of the best books I've read on AI."
— ADAM GRANT, *New York Times* bestselling author of *Originals* and *Give and Take*, and host of the chart-topping TED podcast *WorkLife*

"Kanaan's book makes us aware of the urgent need for international understanding and a formal agreement on AI. Without binding commitments, the future will pose threats, both military and social, that risk our very survival. AI may be a blessing, but it can also be the ultimate curse. The world must agree to draw a red line between the two, and make sure that no one crosses it."
— MUHAMMAD YUNUS, Nobel Peace Prize winner, father of microfinance and social business, recipient of US Presidential Medal of Freedom and US Congressional Gold Medal

"Never have I read a book that did a better job of putting the challenges and prospects of artificial intelligence into context. It's an exceptionally rich context, involving science, history, technology, and our current geopolitical situation. Reading this book will not only help you understand what AI is all about, it will help you understand how it fits into the world today and into the future."
— SEAN CARROLL, theoretical physicist at the California Institute of Technology, host of *Mindscape* podcast, and *New York Times* bestselling author of *Something Deeply Hidden*

"Kanaan recounts the history of AI and why its rapid advance prompts both hopes and fears. He offers a valuable and distinctive perspective on the international tensions it may create."
— MARTIN REES, UK Astronomer Royal and former president of the Royal Society of London

"*T-Minus AI* is enormously illuminating—a fascinating deep dive into one of the most important issues of our day by a leading expert in the field—who also happens to be a riveting writer. I haven't learned so much from a book in ages."
— AMY CHUA, Yale Law professor and *New York Times* bestselling author of *Battle Hymn of the Tiger Mother* and *Political Tribes: Group Instinct and the Fate of Nations*

"Mike Kanaan is a driving new voice in the field of AI. His explanations, insights, and perspectives are trustworthy and brutally intelligent. He's the voice of reason in the room, the one to look to."
—JEREMY BASH, NBC News analyst and
former chief of staff for the CIA and Department of Defense

"As a tech venture capitalist, I know how critical it is that people understand artificial intelligence. *T-Minus AI* explains what we all need to know—not only what AI is, but also the great implications going forward. Eye opening and extremely entertaining."
—JOE MONTANA, tech investor, founding partner of Liquid 2 Ventures,
four-time Super Bowl champion, and NFL Hall of Fame quarterback

"If we are going to prepare our students to lead the next generation, every educator needs to understand the impact of AI on our society. Kanaan's work is the perfect resource to bring you up to speed and to understand the history, scope, and future of AI technology. Leaders and teachers at all levels of education, along with their students, need this book on their reading lists."
—JEFF CHARBONNEAU, former US National Teacher of
the Year, and former finalist for the Global Teacher Prize

"*T-Minus AI* is a must-read about the technology that will drive massive social and political change in the future. If you want to understand AI and its impending impact on the world, this is the book to read!"
—JORDAN HARBINGER, creator and host of *The Jordan Harbinger Show*

"*T-Minus AI* is a thought-leading gem that explains the underlying technology and geopolitical power of AI. Brilliantly written, immensely informative, and as entertaining as a great novel!"
—AUGUST COLE, coauthor of *Ghost Fleet: A Novel of the Next
World War* and *Burn-In: A Novel of the Real Robot Revolution*

"For an accessible and sober explanation of today's most transformative technology, this is THE book to read. *T-Minus AI* will be part of a new canon."
—KARA FREDERICK, technology and national security fellow
at Center for a New American Security (CNAS)

"Part explainer. Part call-to-arms. Kanaan's book makes AI understandable and with that it also makes something clear: AI is both a threat and opportunity. A perfect hybrid of educational and thrilling, *T-Minus AI* is an essential read."
—ALLEN GANNETT, author of *The Creative Curve*

T-MINUS
AI

T-MINUS
AI

Humanity's Countdown to Artificial Intelligence
and the New Pursuit of Global Power

MICHAEL KANAAN

BenBella Books, Inc.
Dallas, Texas

Figure 4.1 A and B: National Archives. Figure 4.2 A: Bob Lord / Bob Lord's Crypto Museum / CC BY-NC 4.0. Figure 7.1: Library Company of Philadelphia. Figure 7.2: white and black stones only, User:Dashed / Wikimedia Commons / CC BY-SA 3.0. Figure 11.1 B: photograph of Atlas® robot courtesy of Boston Dynamics, Inc; C: NASA; D: Photograph of Kilobot courtesy of K-Team Corporation.

BenBella Books, Inc.
10440 N. Central Expressway, Suite 800
Dallas, TX 75231
www.benbellabooks.com
Send feedback to feedback@benbellabooks.com

BenBella is a federally registered trademark.

Printed in the United States of America
10 9 8 7 6 5 4 3 2 1

Library of Congress Control Number: 2020008961
ISBN 978-1-948836-94-4 (trade cloth)
ISBN 978-1-950665-13-6 (electronic)

Editing by Keith Mansfield
Copyediting by Scott Calamar
Proofreading by James Fraleigh and Jennifer Greenstein
Indexing by Debra Bowman
Text design and composition by PerfecType, Nashville, TN
Cover design by Pete Garceau
Printed by Lake Book Manufacturing

Distributed to the trade by Two Rivers Distribution, an Ingram brand
www.tworiversdistribution.com

Special discounts for bulk sales are available. Please contact bulkorders@benbellabooks.com.

We would be better to see in the dark,
than to stumble our way through.

CONTENTS

AUTHOR'S NOTE

The countdown to artificial intelligence (AI) is over. Early AI and machine learning applications have launched from platforms across the globe, and they are already influencing our lives in ways far greater than most people realize.

This book isn't meant to be an excited portrait of a utopian future, nor a dark and dystopian sketch of all we could fear. It is instead offered as an explanation of an incredible evolution in technology, and of a resulting capability that will forever change our information, opportunities, and interactions.

How well we understand the rudiments and real potential of AI, anticipate its social and geopolitical implications, and coordinate the course ahead are imperative matters. Conversation is critical. But, for the dialogue to be meaningful, we must have a common understanding of AI, a common appreciation of its potential, and a common recognition that, like all powerful tools, it will be put to use by organizations and nations with distinctly different agendas and for ideologically opposed purposes.

Some applications of AI will be consistent with Western standards and expectations. Others will be completely contrary. Some we will consider democratically acceptable. Others will shake the foundations of our societies and undermine our core ways of life. Some we'll be able to insulate and secure ourselves from, but others will infiltrate our institutions through methods and means we might not even detect or perceive.

Our focus now must be to openly address the current realities of AI to ensure, as well as we can, that it is implemented only in ways consistent with fundamental human dignities . . . and only for purposes consistent with democratic ideals, liberties, and laws.

My hope is that this book will help enable and inspire that conversation.

PROLOGUE

○———————○

OUT OF THE DARK

It was a Friday morning, September 1, 2017, and not yet dawn when I stepped out of Reagan National Airport and followed my bag into the back of a waiting SUV. After flying east all night from San Francisco to DC, I still had two hours before a Pentagon briefing with Lieutenant General VeraLinn "Dash" Jamieson. She was the deputy chief of staff for Air Force Intelligence and the country's most senior Air Force intelligence officer, a three-star responsible for a staff of 30,000 and an overall budget of $55 billion.

As the Air Force lead officer for artificial intelligence and machine learning, I'd been reporting directly to Jamieson for over two years. The briefing that morning was to discuss the commitments we'd just received from two of Silicon Valley's most prominent AI companies. After months of collective effort, the new agreements were significant steps forward. They were also crucial proof that the long history of cooperation between the American public and private sectors could reasonably be expected to continue. With the world marching steadfastly into the promising but unsettled fields of AI, it was becoming critical that Americans do so, if not entirely in harmony, then at least to the sounds of the same beat.

My apartment was only a short ride away. I was looking forward to a hot shower and strong coffee. But as the SUV pulled out of the terminal and into the morning darkness, a message alert pinged from my phone. It was a text from the general. Short and to the point, as usual. "See Putin comments re AI."

A quick web search pulled up a quote already posting to news feeds everywhere. At a televised symposium broadcast throughout Russia only an hour earlier, Putin had crafted a sound bite making headlines around the globe. His unambiguous three sentences translated to: "Artificial intelligence is the future, not only for Russia, but for all humankind. It comes with colossal opportunities, but also threats that are difficult to predict. Whoever becomes the leader in this sphere will become the ruler of the world."[1]

As the driver accelerated up the I-395 ramp toward the city, a heavy rain started to fall, hitting hard against the car's metal surfaces. Far off, through the window on my right, the dome of the Capitol Building glistened in white light beyond the blurred, dark space of the Potomac. Playing at background volume over the front speakers, an NPR newscaster was describing a three-mile-wide asteroid named Florence. Streaking past our planet that morning, the massive rock would be little more than four million miles away at its closest point—tremendously far by human standards, but breathtakingly near by the infinite scales of space. It was the largest object NASA had ever tracked to pass so closely by our planet.[2] On only a slightly different trajectory, it would have altered Earth's entire landscape. And, like for the dinosaurs before us, it would have changed everything. It would have changed life. *A perfect metaphor*, I thought, *impeccably timed to coincide with Putin's comments about AI.*

I looked back at his words. The message they carried rang like an alarm I didn't need to hear, but the motivation behind them wasn't so clear. Former KGB officers speak carefully and only for calculated reasons. Putin is no exception. His words matter, always. And so does his purpose. But what was it here? Just to offer a commentary or forecast?

No. Not his style. A call to action, then, to energize his own population? Perhaps. But, more than that, this was a statement to other statesmen, a confirmation that he and his government were awake and aware that a sophisticatedly deep effort was underway to accomplish a new world order.

Only a month earlier, China had released a massive three-part strategy aimed at achieving very clear benchmarks of advances in AI. First, by 2020, China planned to match the highest levels of AI technology and application capabilities in the US or anywhere else in the world. Second, by 2025, they intend to capture a verifiable lead over all countries in the development and production of core AI technologies, including voice- and visual-recognition systems. Last, by 2030, China intends to dominantly lead all countries in *all* aspects and related fields of AI.[3] To be the sole leader, the world's unquestioned and controlling epicenter of AI. Period. That is China's declared national plan.

With the Chinese government's newly published AI agenda available for the world to see, Putin's words resolved any ambiguity about its implication. True to his style, his message was clear and concise. "Whoever becomes the leader . . . will become the ruler of the world."

Straightforward, I thought. *And he's right.* But focused administrations around the globe already know the profound potential of AI. The Chinese clearly do—it's driving their domestic and foreign agendas. And the Saudis, the EU nations, the UK, and the Canadians—they know it too. And private enterprise is certainly focused in, from Google, Facebook, Amazon, Apple, and Microsoft to their Chinese state-controlled counterparts—Baidu, Alibaba, Tencent, and the telecom giant Huawei.

AI technologies have been methodically evolving since the 1960s, but over most of those years, the advances were sporadic and relatively slow. From the earliest days, private funding and government support for AI research ebbed and flowed in direct relation to the successes and failures of the latest predictions and promises. At the lowest points of progress, when little was being accomplished, investment capital dried up. And when it did, efforts slowed. It was the usual interdependent circle of cause and effect. Twice, during the late '70s and then again during the late '80s and early '90s, the pace of progress all but stopped. Those years became known as the AI winters.

But, in the last 10 to 15 years, a number of major breakthroughs, in machine learning in particular, again propelled AI out of the dark and into another invigorated stage. A new momentum emerged, and an unmistakable race started to take shape. Insightful governments and industry leaders began doing everything possible to stay within reach of the lead, positioning themselves for any possible path to the front.

Now, for all to hear, Putin had just declared everything at stake. Without any room for misunderstanding, he equated AI superiority to global supremacy, to a strength akin to economic or even nuclear domination. He said it for public consumption, but it was rife with political purpose. *Whoever becomes the leader in this sphere will become the ruler of the world.*

Those words would undoubtedly add another level of urgency to the day's meetings. That was certain. I redirected the driver to the Pentagon and looked down at my phone to answer the general's text. "Landed. Saw quote. On my way in."

The shower would have to wait.

———

In the months that followed, Putin's now infamous few sentences proved impactful across continents, industries, and governments. His comments provided the additional, final push that accelerated the planet's sense of seriousness about AI and propelled most everyone into a higher gear forward. Public and private enterprises around the globe reassessed their focuses and levels of commitment. Governments and industries that had previously dedicated only minimal percentages of their R&D budgets to the new technology suddenly saw things differently. It quickly became unacceptable to slow-walk AI efforts and protocols, and no longer defensible to incubate AI innovations for longer than the shortest time necessary.

Now, not long after, the pace of the race has quickened to a full sprint. National strategies and demonstratable use have become the measurements that matter. Rollouts have become requisite. To accomplish them, agendas are more focused, aggressive, and well funded. Sooner than many expected, AI is proving itself a dominant force of economic, political, and cultural influence and is poised to transform much of what we know and much of what we do. China, Russia, and others are utilizing AI in ways the world needs to recognize. That's not to say all efforts and iterations in the West are without criticism. They're not. But if this new technology causes or contributes to a shift in power from the West to the East, everyone will be affected. Everything will change.

The future is here, and the world ahead looks far different than ever before.

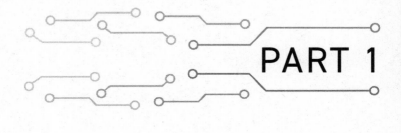

PART 1

THE EVOLUTION OF INTELLIGENCE

From a Bang to a Byte

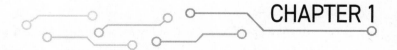

CHAPTER 1

SETTING THE STAGE

Philosophically, intellectually—in every way—human society is unprepared for the rise of artificial intelligence.

—Henry Kissinger
Former US Secretary of State and US National Security Adviser
"How Enlightenment Ends," The Atlantic, *June 2018*

No longer just science fiction or fantastic speculation, artificial intelligence (AI) is real. It's here, all around us, and it has already become an integral and influential part of our lives. Although we've taken only our first few steps into this new frontier of technological innovation, AI is providing us powerful new methods of conducting our affairs and accomplishing our goals. We use these new tools every day, usually without choice and often without even realizing it—from applications that streamline our personal lives and social activities to business programs and practices that enable new ways of acquiring competitive advantage. At home and recreation, our methods and mechanisms are changing. At work, the same — our operational programs, production processes, marketing strategies, sales efficiencies, and service systems are all evolving. In research, experts across all fields are now using AI to ask deeper questions and obtain previously unavailable insights and answers. Governments around the globe are increasingly using AI technologies to advance their national agendas and to protect the cultural integrity, assets, and security of their citizens. In short, the individuals, organizations, and nations at the forefront of AI are learning to wield tremendous new advantages over those that lag behind. But, even for those at or near the lead, there are significant risks lurking in the days ahead. And for those that fail to keep pace, there will be substantial costs and losses to suffer. In the age of artificial intelligence, second place will be of an ever-diminishing and distant value.

Every day, I speak with people from various parts of the private and public sectors — including leading technology firms, academic institutions, national research laboratories, and policy think tanks. I've been fortunate to advise US congressional members, brief American and foreign military personnel, and give presentations to audiences from a wide variety of backgrounds, educations, and occupations. Through it all, I've learned a lot about the common misperceptions and misgivings people have when trying to understand AI. Most conversations about artificial intelligence, whether in auditoriums, offices, or coffee shops, either begin or end with one or more of the following questions:

1. What exactly *is* AI?
2. What aspects of our lives will be changed by it?
3. Which of those changes will be beneficial and which of them harmful?
4. Where do the nations of the world stand in relation to one another, especially China and Russia?

5. And, what can we do to ensure that AI is only used in legal, moral, and ethical ways?

Although the answers to those questions merit long discussions and are open to differing opinions, they should at least be manageable and factually accurate. The topics shouldn't be too difficult to discuss or debate—not conversationally or even at policy-making or political levels. Unfortunately, they generally are.

But the conversational disconnects that usually occur aren't because of some complex technical details or confusing computer issues. Instead, it's usually, simply, because of the same old obstacles that too often stand in the way of many other conversations. Regardless of the topic, and even when it matters most, we too frequently speak below, above, around, or past one another—especially when we don't have an equal amount of information, a shared base of knowledge, or a common set of experiences. In those instances, we make too many assumptions, allow too many things to go without saying, and use too many words that hold different meanings for different people. In short, too many confusions are never clarified and too many more are created. As a consequence, we're doomed for frustration and failure from the start, inevitably unable to understand one another and incapable of appreciating each other's perspectives and talking points. My goal throughout this book is to avoid those pitfalls.

The best way to start is to first address the most common misperceptions of all, the ones we tend to bring with us *into* the AI conversation. The first of these is the assumption that AI is unavoidably destined, sooner or later, to develop its own consciousness and its own autonomous, evil intent. For that idea, we can thank science fiction and the entertainment industry. Make no mistake, I'm an ardent fan of science fiction, both on-screen and in books. Without any doubt, the sci-fi genre has given us fine works of imagination, insight, and art. Many great fiction writers and filmmakers are extremely knowledgeable about technology and conscientiously concerned about our future. Time and again they've proven themselves true visionaries, and we're unquestionably better off for their work. They spark our curiosity, ignite our imaginations, increase our appetite for knowledge, and encourage our interests in science and societal issues.

But, when it comes to their scientific portrayals of artificial intelligence, our most popular authors and screenwriters have too often generated an array of exotic fears by focusing our attention on distant, dystopian possibilities instead of present-day realities. Science fiction that depicts AI usually aligns a computer's intelligence with consciousness, and then frightens us by portraying future worlds in which AI isn't only conscious, but also evil-minded and intent, self-motivated even, to overtake and destroy us. To create drama, there has to be conflict, and the humans in these stories are almost always overwhelmed and outmatched, naturally unable to compete against the machines' vastly superior intelligence and mechanical strength. Iconic movies like *2001: A Space Odyssey*, *The Matrix*, *The Terminator*, *Ex Machina*, and *I, Robot*, along with television series such as *Westworld* and *Black Mirror*, have turned our underlying fears and suspicions into deep-seated and bleak expectations.

Even today, commercial companies that offer AI products and consumer services routinely have to fight our distrust of intelligent machines as a basic, necessary part of their regular marketing efforts. Just think of all the television commercials for AI-enabled products we now see, and consider how many of them are focused first on trying to put us at ease by casting a polite and gentle glow to the figurative, artificial *face* of their AI, even when that face has absolutely nothing to do with the services their products actually provide. By the end of this book, I'm hopeful that even the most pessimistic among you will come to agree that conscious, evil-inclined AI is not the unavoidable endgame, nor the thing we have most to fear. At the end of it all, and as has always been the case, *people*—and the specific uses to which we ourselves put our machines—are and will remain the principal problem. It's what *we* will do with AI that matters . . . and, yes, the potential for human misuse, intentional and otherwise, is worthy of great concern.

AI is an extremely powerful tool, and it has immense implications we must consider and evaluate carefully. It's a very sharp instrument that shouldn't be callously wielded or casually accepted, especially when it's in the wrong hands or when it's used for intentionally intrusive or oppressive purposes. These are serious issues, and there are significant steps we must take to ensure AI is properly designed and implemented. Fortunately, and contrary to what many people think, it's not necessary to have a background in computer science, mathematics, or engineering in order to very meaningfully understand AI and its technological implications. With just a basic comprehension of a few fundamental concepts behind today's computers and related sciences, it's entirely possible to connect the relevant dots and understand the overall picture.

As we'll see, creating tools to facilitate our lives is the strength of humankind. It's what we do. Given enough time, it was arguable, perhaps even inevitable, that we would create the ultimate tool—artificial *intelligence* itself. But, what exactly does it mean that we've accomplished that task? And how is AI even possible? In large part, the answers lie in the history of ourselves and of our own biological intelligence. It turns out that artificially replicating what we know about the human thought process, at least as best we can, is a highly effective blueprint for creating something similar in a machine. With that in mind, let's draw back the curtain on the human experience—because it's our own evolution and our own history, that teaches us the fundamentals that make it all possible.

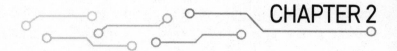

CHAPTER 2

IN THE BEGINNING . . .

For millions of years, mankind lived just like the animals. Then something happened which unleashed the power of our imagination. We learned to talk and we learned to listen.

—*Stephen Hawking, 1942–2018*
Theoretical Physicist and Cosmologist

We know of only one example of advanced intelligence in the universe—and that's our own. To best understand *artificial* intelligence and appreciate its impending role for the future, it's helpful to briefly address some fundamental realities about the gradual evolution of humanity's biological place on the planet, about *human* intelligence, and about our unique ability to share information and collectively learn. The story is best told if we start . . . at the start.

———

In the flash of an instant some 13.8 billion years ago, all space, time, energy, and fundamental forces in the known universe burst into existence. As the energy slowly cooled, the earliest of all matter formed in relatively equal measures throughout the expanding universe. Over the next three or four hundred million years, gravity slowly bound some of the matter together and the first generation of stars began to form. Far larger and hotter than today's stars, they burned for just a few million years before they exhausted their nuclear fuels and died.[1] Many, just like today, ended their lives in colossal supernova explosions that returned their spent star stuff back to space, seeding the universe with new, heavier materials for the next generation of stars. Our sun is one of the beneficiaries of those that came before it. Some 4.6 billion years ago, it formed in the outer reaches of an otherwise ordinary galaxy we now call the Milky Way.

A few hundred million years after the sun's inception, an upstart solar system of eight planets formed from leftover stellar debris the sun hadn't taken for itself. One of the planets, Earth, fell into a gravitational orbit 93 million miles out. It was the third planet from the sun. For its first few hundred million years, Earth was a violent world of molten-hot liquid elements, unimaginably hostile and intolerant of any conceivable life. Slowly, though, the planet stabilized, cooled, and crusted sufficiently for solid land and the earliest oceans to appear. But, at Earth's center, an inner cauldron continued to burn that spewed chemical nutrients into the oceans for another billion years. Then, ignited by a mysterious spark we may never understand, carbon-based biological life somehow percolated into a watery existence.

Even after the earliest life began some 3.5 billion or more years ago, the next two billion years brought nothing more than water-bound, single-celled organisms that weren't much different than today's microscopic bacteria. While it's possible that life on the planet could have stayed simple and small, the minuscule organisms thriving throughout the primordial soup of Earth's oceans ultimately developed the ability to group together into more complex, multicellular organisms. From there, cell specialization began. And after another billion years or so, the planet's oceans and atmosphere eventually filled with oxygen, allowing life to drastically change its size and shape. Complex species of microscopic marine arthropods developed and, eventually, small fish evolved. Energized approximately 2.4 billion years ago by increasing food supplies during what's called the Great Oxidation Event,[2] life in the oceans gradually grew larger and larger. Land plants appeared, then trees, and then entire forests. Finally, the earliest amphibians slithered out of the water and crawled onto land with fins that slowly turned into limbs.[3] This was the Devonian period, only about 400 million years ago.[4]

Evolution on this grand of a scale had been slow, incomprehensibly so from a human perspective. Significant biological change doesn't occur in years or decades—as we'll later see that it does with respect to life-altering modern technology—but only in thousands, millions, and billions of years. From the time life first arose, it had taken more than three billion years for the first true land animals to appear. And, even when grounded solidly on soil, life still continued to evolve on a necessarily slow, natural pace. After another 150 million years, though, a highly adaptable and comparatively intelligent new class of animals evolved to take center stage. They were the dinosaurs.

First appearing 235 million years ago, the dinosaurs were an incredibly diverse group of animals that ruled our planet for 170 million years. Over three distinct geologic time periods—the Triassic, Jurassic, and Cretaceous—they survived through drastically different environmental conditions and planetary landscapes. Various species of dinosaurs came and went as they adapted to and dominated every changing niche that nature offered, altered, and then abolished. The dinosaurs' diversity ranged from small insect hunters the size of rabbits that weighed only a few pounds to giants over a hundred feet long that carried 70 tons of flesh on their frames. There were carnivores, herbivores, and omnivores. There were hunters, grazers, and scavengers. Some lived solitary territorial lives, and others roamed in cooperative packs and large herds. Some were horned, some armored, others scaled or feathered. Most walked only on land, but some waded in shallow waters and others returned to live their entire lives in the seas. Some took to the air and flew. Contrary to traditional thought, some may even have been warm-blooded, like mammals.[5]

In light of the dinosaurs' incredible range of adaptability and dominance over the natural and competitive environments of their time, it's entirely possible their reign could have continued, even until today, if not for a single, remarkable stroke of misfortune. Sixty-five million years ago, a massive meteor struck Earth with so much force that the disruption it caused extinguished the dinosaurs from the planet. The meteor was over seven miles wide, larger than Mount Everest, and weighed more than 100 billion tons. It sped through Earth's atmosphere at almost 15 miles per second before slamming into the planet along what is now the Yucatán Peninsula in the Gulf of Mexico. Due to its sheer size and speed, it blew a hole in the Earth almost 20 miles deep and 124 miles wide. The impact released more energy than a hundred teratons of TNT, almost ten billion times the power of the atomic bomb detonated over Hiroshima.[6] Plate tectonics were shaken, seismic tremors and cataclysmic quakes rocked the planet. Massive tsunamis and tidal waves raged from the oceans onto land. Volcanoes and fissures blew open and spewed lava over the landscape. Steam, sulfur, and ash bellowed into the skies. So much residue was thrust into the atmosphere that a hailstorm of thousands more meteors streaked back into and through the oxygen-rich skies, the combined heat of which literally set the air on fire.

When the turmoil subsided, a dark and cold winter fell over the planet. Plant life that hadn't burned off in the direct aftermath of the meteor's impact was soon choked out from the light. Death led to more death throughout the food chain, as animals that hadn't immediately been killed eventually died of starvation or exposure to colder temperatures, acid rain, and noxious chemicals that seeped from the ground and permeated the air for years, if not decades. In only a short period of time—far less than a comparative blink of an eye—170 million years of successful adaptation and planetary dominance was rendered fully extinct. Done. The dinosaurs, other than a few avian survivors from which all of today's bird species descend,[7] were gone. So too were up to 70 percent of all other species then on the planet. The landscape was decimated and the order of evolution that had prevailed, so powerfully and for so long, was suddenly stopped in its tracks.

If not for the meteor and the mass extinction it caused, no one can say how long the dinosaurs may have continued to thrive, or into what creatures any of their species may have evolved. Likewise, no one can definitively say what would have become, or not, of the other species that slowly emerged in their absence. For anyone questioning the unpredictability of evolution, just consider that all of today's whales evolved from a four-legged land animal that lived long after the dinosaurs and looked somewhat like a wolf.[8] In any event, we humans are fortunate the meteor struck. We owe our very existence to it.

———

The catastrophe that wiped out the dinosaurs created new opportunities for the few classes and species of life that managed to survive. Those that benefited most were the lowly ground-burrowing mammals. Prior to the meteor, they'd lived their lives mostly underground to avoid the constant predatory menace of the dinosaurs. But with the dinosaurs no longer a threat, they soon spread over the landscape and began to flourish throughout the ecological niches the dinosaurs had vacated. For the first time, these small, previously inconsequential animals were able to compete for and obtain dominant roles that were never before available. Over the 60 million years following the dinosaur extinction, the warm-blooded mammals rapidly—at least by evolutionary scales of time—increased in diversity and size. They established strong new footholds across the entire planet and ultimately occupied ecological spaces throughout all of the deserts, forests, savannas, and wetlands as they branched out in the ever-growing evolutionary tree of life.

Then, only five to six million years ago, the first of the hominid family of mammals appeared . . . and, in them, came the possibility for our eventual human form of biological complexity. Although it would still take millions of years more for the earliest humans to evolve, these early hominids were our most ancient, direct ancestors. They stood distinct from the other primates. And, although they still had wildly different anatomical traits from those that would much later distinguish the first humans, the earliest hominids nonetheless began the evolutionary path that eventually led to ourselves.[9]

———

Beginning about two million years ago, the earliest humans—*Homo habilis* and *Homo erectus*—emerged at last in Africa and then gradually made their way to parts of prehistoric Europe, the Middle East, and Asia. At least four other now extinct species of humans later appeared— *Homo rudolfensis*, *Homo heidelbergensis*, *Homo denisova*, and *Homo neanderthalensis*. The last of them, the Denisovans and the Neanderthals, died out roughly 50,000 to 30,000 years ago, respectively.[10]

The most recent human species to appear was our own, a mere 200,000 years ago. In the greater timeline of the evolution of life on this planet, that was less than a finger snap ago. From life's earliest appearance, it had taken 3.5 billion years or more for the specific biological complexities of modern man to finally emerge. Anatomically, what distinguishes us most significantly from our human predecessors is our enlarged neocortex, which is the outermost layer of our brain that manages our higher-order functions like sensory perception, cognition, motor commands, spatial reasoning, and language. The neocortex was the biological enhancement that gave us the eventual combination of intelligence factors necessary to rise far above all other animal life . . . and, ultimately, to consider, comprehend, and manipulate the very essence of nature itself. As the most modern and only surviving species of humans, we've aptly named ourselves *Homo sapiens*, which, from Latin, roughly translates to "wise man."[11]

With the higher intellectual abilities provided by our enhanced brain structure, our first great breakthrough in realizing our intellectual potential came when we started to develop language—a foundational construction, as we'll later explain, that would one day prove critical even for artificial intelligence. It's believed that we first began speaking with a modern verbal range sometime around 100,000 years ago.[12] We may never know whether language began independently in several different places and times, or if it initiated in a single place and branched out from there. Either way, the development of verbal language was a tremendous step forward for our species. It allowed us to communicate locally and gave us the ability to efficiently cooperate in our mutual efforts to better understand, manage, and survive the world around us.

By 50,000 years ago, we were well into the process of developing our verbal languages. We continued to migrate across the landscape and slowly learned to manipulate nature to our increasing advantage. But still, as the most intelligent species to ever inhabit the planet, we essentially did little more for the next 35,000 years except wander around, nomadically, with our crudely formed stone weapons and primitive tools, habitually following food sources and building campfires to cook meat for our small clans, to keep us warm, and to fend off wild animals.

Eventually, though, we started to settle down by building long-term communities with habitats and structures of lasting permanence. Even then, we didn't grow our own food or domesticate animals until 10,000 years ago. It took another millennia or two to put metals to use only 8,000 to 9,000 years ago. And we didn't form civilizations that included specialized social and community roles until 6,000 to 7,000 years ago.

To put the slow pace of our intellectual innovations into perspective, we didn't invent the wheel until 5,000 to 6,000 years ago. Think about that. Over the entire timeline of *Homo sapiens'* existence, and as the most advanced and only remaining of all human species, it took 194,000 of our 200,000 total years on Earth to finally piece together the idea and method of putting a round object to a constructive, locomotive use. How imperceptive and incapable of combining and processing mutual thoughts and experiences we must have been for it to take a full 97 percent of our species' evolutionary timeline just to innovate a mechanical means of rolling something rather than pushing, pulling, or carrying it.

About the time that the wheel was coming into common use in the earliest of our civilizations, so too was another, even more significant innovation. A little more than 5,000 years ago, the ancient Sumerians (in modern day Iraq) first began reducing their verbal language to writing.[13] This was a tremendously important step forward in the use of language—it enabled true collective learning.

The concept of recording and communicating events and ideas in ways more permanent than verbal exchanges first appeared as pictographs (pictures that were drawn or inscribed to represent objects in a descriptive format). Then ideographs that represented ideas and thoughts, rather than objects, emerged as more useful and flexible versions of pictographs. Eventually, logographs and hieroglyphics that represented full words taken directly from verbal languages were created.

From logograph and hieroglyph symbols that represented full words, the next step in written language was the development of letters that represented voice sounds. Sometime between 1850 and 1700 BCE, along the banks of the Mediterranean Sea in the region of present-day Lebanon, letter symbols representing the full range of independent vowel and consonant sounds were created and combined for a true written alphabetic language that represented all of the phonetic sounds of the underlying verbal language. This was the proto-Canaanite alphabet, also known as the proto-Sinaitic alphabet. It consisted of 22 letters, each of which represented a single sound of voice, or phoneme.[14]

The proto-Canaanite alphabet was a concept far removed from the earlier symbol-based writings. With alphabet-based writing, our collective ability to efficiently learn from one another was finally unleashed. Now, having a completely expressive written language, we were able to record and preserve information, experiences, ideas, and observations with the nuances and sophistication of verbal language, and we were able to do it with a universality that could efficiently be passed down generationally and cross-culturally. Rather than requiring thousands of distinct culture-based symbols or complex characters for each word or concept, as do pictographic and logographic systems of writing, an alphabet-based writing system only requires one letter (or a small combination of letters) for each distinct sound of speech in the verbal language. Accordingly, alphabet-based systems only have to contain enough characters so that, alone or in combination, they can represent all standard sounds in the language.

While languages differ in the specific sounds of their phonemes, the total number of them in any one language is relatively small. For instance, depending on the dialect, English has approximately 44 phonemes, French about 37, and Spanish only 25.[15] With this greater simplicity over pictograph or logograph systems came a number of distinct advantages. First, an alphabet-based written language is much easier to teach and to learn, since it is easier to memorize and manipulate dozens of characters than it is to remember hundreds or thousands of symbols. Also, once learned, an alphabet-based written language makes mastering the verbal language upon which it is based likewise easier to accomplish. As a result of both of those advantages, overall literacy and language proficiencies in alphabet-based societies became much easier to achieve and thus made generational exchanges of information more efficient.

With experiences, thoughts, and ideas more easily communicated through manageable writing systems, modern humans could finally share information other than verbally or visually. Previously, our ability to teach and to learn had been limited to our direct contact with one another. Time, access, and other realities being what they were, there was only so much information any one person could ever individually acquire through his or her own experience or direct contact with others. Now, though, with efficient and expressive writing systems, we could learn *collectively*. We could record information, share it, accumulate it, compare it, and analyze it.

From the moment alphabet-based written languages took hold and began spreading throughout different civilizations, a new game was truly on. Human learning, human capability, and human dominion over nature began to expand at an unprecedented rate. For the next few thousand years, we utilized this new catapult to share information and solidify the rudiments of our most basic and important knowledge bases. Then, in the mid-fifteenth century, around 1440, the collective value of written language became even greater when a German blacksmith named Johannes Gutenberg invented a printing press with a valuable adaptation borrowed from the East: moveable type.[16] Prior to Gutenberg's invention, books were copied almost exclusively by hand, which was a laborious process that usually took months to complete. As a result, longer works were extremely expensive and generally available only to the rich and educated upper classes. The printing press, however, had an immediate and dramatic effect. The efficiency it brought to the copying process allowed cumulative information in the form of books to spread quickly and affordably. By the turn of the sixteenth century, the Gutenberg press was common throughout all of Western Europe and other parts of the world. The era of widespread, efficient communication had arrived. Information had become available to the masses, and with it, the potential of humanity changed forever.

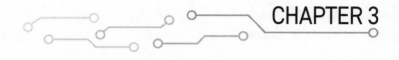

CHAPTER 3

TOO MANY NUMBERS TO COUNT

Man is equally incapable of seeing the nothingness from which he emerges and the infinity in which he is engulfed.

—Blaise Pascal, 1623–1662
French Mathematician, Physicist, and Inventor

J ust as the development of verbal and written languages allowed humans to efficiently collect, analyze, and share narrative information, the innovation of numbers empowered us to effectively measure, calculate, and exchange mathematical information. Most aspects of our world, particularly our physical world, are best described by numbers and by the mathematical rules we later derived from them. Combined, numbers and math give us the ability to assess the basic components of nature and to evaluate the most fundamental things around us—quantities, properties, distances, weights, and time.

To help understand some of the most important characteristics of AI, it's important to first conceptually appreciate the vast sizes and scales of extremely large numbers—the kind of numbers we don't encounter in our everyday lives, but that facilitate and drive the fundamentals of modern computing, artificial intelligence, and machine learning. As humans, we're ill-equipped to easily comprehend just how big numbers can be. We don't have any natural feel for the enormities represented by the most extreme quantities, sizes, and speeds. None of us do. This isn't because of any deficits in our individual IQs or levels of education. It's simply because of more fundamental, ingrained human realities—it's because of who and what we are as a result of our historical circumstances and experiences . . . again, our biological and social evolution.

Until very recently, extreme numbers were of no relevance to us. For almost all of our species' time on the planet, there was never any practical need for humans to consider, much less fully comprehend, any numbers beyond the most relatively small. Large numbers initially contributed nothing to our perception of the world or to our ability to manage our way through it. As early humans, we wouldn't have gained any competitive survival advantage from being able to comprehend what a million or a billion of anything was. Frankly, for the vast majority of our history, we didn't even need to understand what a hundred or a thousand of anything was. We just needed to recognize and manage the small numbers and quantities that mattered to us on a daily basis—like the number of members in our clan, the number of wolves outside our cave, the number of logs necessary to maintain our fire, or the number of mammoths we had to separate from the herd in order to have a reasonable hunter's chance.

As time went on, however, large numbers gradually became more and more relevant to us. The sizes of our communities expanded, the distances we traveled increased, and the number of warriors and weapons in the armies of our enemies grew. Still, though, until only recently in the timeline of our evolutionary history, we had no use for the kinds of numbers we now routinely use—like million, billion, and trillion. As a result, our brains don't intuitively understand them, for we don't have any evolved, natural sense of the immensities they represent.

It wasn't so long ago that the number of stars we see in the sky was considered an unimaginable quantity, even though on a clear night a human can only see a few thousand with the naked eye. Now, though, we look up to the night sky and wonder in amazement at the hundreds of billions of stars we've learned exist in our galaxy, at the hundreds of billions of galaxies in the known universe, and at the incredible speeds at which things in the universe, like light, travel. But, even so, most of us have no real context for things so numerous or so fast. It's just too much to wrap our heads around.

But even the digital devices and personal computers we now routinely purchase contain circuit boards with billions of micro transistors that *we* build into them. Moreover, those billions of micro transistors can analyze billions of bits of information and perform hundreds of billions of calculations per second. In fact, our fastest supercomputers can now perform 200 quadrillion (which is 200 with 15 zeros behind it) calculations in a single second.[1] But unless we engage in science, computer technology, or mathematics every day, or unless we have some professional need or intellectual compulsion to comprehend the meaning of such astoundingly large numbers, it's doubtful we have any accurate notion of just how big a billion or a trillion truly is, let alone a quadrillion.

We certainly know a million is a lot, but what does "a lot" really mean? And how much more is a billion? A trillion? It's easy to say that a thousand multiplied 1,000 times is a million, that a million then multiplied 1,000 times more is a billion, that a billion then multiplied yet another 1,000 times is a trillion, and so on. But that sentence is just a mathematical description of incomprehensibly big numbers by using multipliers to arrive at other outrageously big numbers. It doesn't really help. And although scientists use prefixes such as *kilo, mega, giga, tera,* and *peta* to indicate successive multiplication by factors of a thousand (and *milli, micro*, *nano*, *pico*, and *femto* for division by successive thousands), their terminology doesn't help our appreciation of the quantities either.

A better way to understand the absurd amounts these numbers represent, and the tremendous spans of differences that lie between them, is to frame things in nonmathematical contexts with which we're more naturally familiar. Time and distance are two of the best ways of doing so. A common thought experiment using the concept of time is to imagine counting to the number 1,000—not an unreasonable task, although I suspect few of us have actually done it. In any event, if you start with the number one and count one additional number every second, without stopping, it will take you almost 17 minutes to reach 1,000. That's not tough to calculate. Just divide 1,000 by 60 (the number of seconds in a minute). And while it might take a bit longer than you'd have guessed, it's easy to see, and you're probably thinking, *"OK, sure, that sounds about right."*

But what if you want to continue counting from one thousand to one million? How much longer would that take? This time, the answer's more likely to surprise you. If you continue counting, again by adding one new number every second and again without stopping (not even to speak, eat, drink, or sleep), it will take more than 277 hours, or more than 11½ continuous, uninterrupted days. And if you want to be more realistic in going about the task, by only doing your counting during the course of your eight-hour workdays, then the job of counting to a million would take you almost two months of a full-time work schedule to complete—and that's without time off for lunch.

And what about a billion? That's a number we now hear all the time. It can't be *that* much more than a million, right? Wrong. If you want to count to one billion at one-second intervals, you'll unfortunately have to spend most of your adult life at the task, because it will take almost 32 years of continuous, nonstop counting to get there. And to count to one trillion, another number that we're beginning to hear and use more frequently? Well, that would take you almost 32,000 years.

Another common version of this thought experiment uses distance to put things more clearly in perspective. Millimeters are relatively small. There are 25.4 of them in an inch. That means that the length of 1,000 millimeters is easy to visualize—it's about 39 inches, not much longer than the length of an average adult stride. Therefore, without even hurrying, you can walk a distance of one million millimeters, or about 1,000 steps, in less than 15 minutes. So far, so good. But, to cover the distance of one billion millimeters you'd have to keep walking, nonstop, for more than a week—because that's how long it would take to walk 1,000 steps 1,000 times. And to walk one trillion millimeters, you would have to travel around the equator, the full circumference of Earth, more than 25 times.

As we get deeper into the AI conversation, we'll discuss numbers that have real meaning and real significance to computer technology that make numbers like million, billion, and trillion absolutely minuscule by comparison. For now, though, it's important just to begin opening your mind to the notion and enormity of extremely large numbers—in the world of modern computing and machine learning applications, they allow for extreme calculations . . . and extreme possibilities.

———

Historically, before large numbers, quantities, and speeds could possibly be understood, in fact before any mathematical descriptions of the world could even be determined, the numbers themselves had to be created. Our ancestors' earliest use of a counting system most surely predated their earliest verbal languages, likely beginning with very simple, primitive processes of grouping small numbers of pebbles, stones, or twigs together to symbolize the number of animals they were hunting or the numbers of members in a rival tribe. It's easy to imagine, for instance, early cave dwellers crouching before their fellow hunter-gatherers and arranging stones together on the ground to represent the

number of wild boar just over the nearest hill, or the number of rival
hunters on the other side of a valley.

But, as time went on, our ancestors eventually realized that it would
be more helpful to use marks or symbols that could be recorded and
preserved. The earliest artifacts representing a recorded numbering system are bones with tally marks notched into them. Found in parts of
Africa and Europe, these Ishango bones — or *tally sticks* as they're commonly known — date back more than 20,000 years and show collections
of notched slashes or strokes, each of which presumably represented
one unit of something. The notched slashes could have been used as a
permanent recording of a one-to-one representation, or they might even
have been used as the earliest of all calculation devices.[2] Either way, these
bones show the formation of numbers at the earliest origins of mathematics. They were collections of strokes, symbols of ones, used to individually and collectively represent the number of items described. This,
the simplest of all counting methods, is referred to as a unary system.

On occasion, we still use tally marks today. The problem with any
tally system, however, is that it uses only a single symbol, an unchanging
slash, to represent every individual unit of items being counted. Consequently, the system is relatively unmanageable for large quantities. A
solution, while still remaining within a single-symbol tally system, is to
arrange large collections of tally marks into smaller, more visually manageable sets. The most common means of doing so is to use every fifth
stroke to strike through the preceding set of four. At a glance, it's then
easy to recognize each set as a collection of five units.

Tallies seem to have been the only systematic method of counting
until 4,000 to 5,000 years ago when the Babylonians created the first
"positional" numbering format. It was part of their cuneiform writing system.[3] A positional numbering format means that the value of a
numeric symbol is dependent upon where it appears relative to another
symbol in the system — either to the left or to the right. Babylonian
cuneiform only used two different number symbols — one of which was
used for each additional unit between one and nine, and the other of

which was used to represent quantities of 10. The system had 59 different groupings of the two symbols to represent the 1 through 59 units. At 60, and with no zero symbol in the system, cuneiform repeats the pattern all over again. Although the cuneiform, base-60 numeral system is rather complicated to use for mathematical purposes, it did have a number of useful aspects and is still used in certain measurement applications today. Our present-day time system, in which hours are broken into 60-minute segments and minutes are then broken into smaller 60-second segments, results from the original cuneiform. So does our description of the angles within a 360° circle, where halfway around the circle is equal to 180° (three sections of 60°), and the full sweep of the circle is equal to 360° (six sections of 60°).

While the Babylonian base-60 positional system remained in use throughout the Egyptian region for thousands of years, different number systems were slowly developing in other locations. The Chinese created a system of patterned slashes they could arrange positionally. It eventually led to the Chinese abacus, which allowed for quick mathematical calculations, primarily by counting in groups of five and ten.

The Greeks, on the other hand, created a number system that used the 24 letters of their written alphabet to stand for progressively growing numbers. For numbers larger than 24, they had to create additional symbols—but as a practical matter of memory and its limitations, they could only employ a limited number of extra symbols. As a result, the Greek system wasn't much good for counting, adding, or recording large numbers.

The Romans also developed a system that utilized letters from their alphabet. But rather than being a positional system, Roman numerals are additive—meaning that a symbol's position establishes whether its value should be added to or subtracted from the letter (numeral) next to it. Most of us are probably familiar with Roman numerals, since the system still makes appearances today—not as a counting or mathematical system, but more for branding or distinguishing purposes. Super Bowl XLV, as an example, signifies the 45th annual occurrence of the game, where 10 (X) is subtracted from 50 (L) before adding 5 (V) to equal 45. And if you turn back to the opening of this book, you'll see that all of the pages before Chapter 1 use Roman numerals as page numbers—which is commonly done in books to distinguish the preliminary pages from those that make up the main body of the work.

———

Speaking of what matters most, the numeral system that we currently use is the Hindu-Arabic system, commonly called Arabic numerals. It's a base-10 decimal system that originated in India sometime during the sixth or seventh century. By the end of the fifteenth century, the system had spread throughout much of the world and was commonly used throughout Europe—aided in great part by the printing press, the system's ease of use, and the numerous advantages it has over any other system used before it.[4] We're all intimately familiar with the Arabic base-10 system because we use it every day. We know that there are only 10 unique numerical symbols in the entire system. Those 10 symbols (1–9 and 0), when used in various combinations and orders, can represent and identify any value necessary. Our system can also express exact values for positive numbers, negative numbers, and decimals—which allows us to represent precise values for the smallest to the largest quantities we can imagine.

Easy to learn and memorize, Arabic numerals are convenient for calculations both simple and complex. The system works well even without being written. When thinking in base 10, we can count and calculate easily, even in our heads—although it doesn't hurt that we have ten fingers to count with when needed. We can estimate the cost of a dinner tab or the approximate total of our grocery list along with any number of other things we regularly encounter. We can comprehend the relevance of a given number of days, months, and years in the course of our lives. We can imagine the size of a wedding reception when told that 300 people will be attending, and, in a comparative sense, we have a clear feel for why it's wrong to drive 50 miles per hour in a 25-mile-per-hour school zone. We can even visualize, somewhat, a crowd of 115,000 fans packed into a football stadium.

All of those examples fall within the scale and scope of our normal experiences or, in the last case, are at least peripherally imaginable to us. But, because we learn and conceptualize new things, even new quantities, by comparing them to things with which we're already familiar, numbers outside the scale of our experience—as we discussed—present real problems for us to comprehend. What's even more difficult to comprehend is how, in a mathematical sense, such numbers can multiply and grow—not just quickly, but enormously . . . *exponentially*. As we'll later see, exponential growth is important to appreciate when discussing not only the capabilities of computers, but also the realities of AI.

There's an old Hindu legend that makes exponential growth quite easy to understand. The story goes like this:

Long ago, a traveling wise man was given the chance to meet with a highly regarded king of a remote region in India. The king was known by his people as the greatest chess player in all the land. As a gift and show of respect, the wise man presented the king with an ornate chessboard made of delicate woods and intricately etched ivory. The king was so grateful for the gift that he challenged the visiting wise man to play him in what would be the very first game on his newly proclaimed royal chessboard. But the wise man declined the offer, respectfully noting that he would be an unworthy opponent, as his skills couldn't possibly match those of the king. Even more impressed by the wise man's humility, the king responded that if the wise man would accept his challenge to play, and was then able to win the game, the king would reward him with anything he wished. With no more hesitation, the wise man accepted the king's challenge. In short order, and to the king's surprise, the wise man easily won the match. The king, it seemed, had been conned.

Suddenly aware that he'd been duped, the king was nonetheless a man of his word. Consistent with his promise, he asked the visitor to name his reward. The wise man replied that all he desired was rice. The king, again impressed by such a modest demand, asked what amount would be sufficient. The wise man replied that he only wanted one grain of rice for the first square of the chessboard, two grains for the second square, four for the third, eight for the fourth, and so on, continuing to double the number of grains for each subsequent square on the board . . . until all 64 of the squares were accounted for. Thinking that request quite reasonable, the king quickly agreed and instructed one of his advisers to pay the debt. The adviser, however, was more astute at mathematical principles than his monarch and immediately realized the extent of the kingdom's dilemma. He quietly explained to the king that neither he, nor all of the other kings on Earth, even combined, could possibly pay the amount the wise man had requested. Although the wise man's request seemed reasonable enough, and although the number of rice grains would start quite small, the rules of exponential growth would quickly escalate the numbers beyond the imaginable.

While the king only owed two grains of rice for the second square and four grains for the third, he would owe 128 grains of rice on the eighth. That was fine, but by continuing to double the amount on each successive square of the chessboard, the numbers would grow so rapidly that for the 21st square alone, the king would owe one million grains. For the 32nd square, two *billion* grains. For the 41st square, one *trillion* grains. And, for the very last of the 64 squares on the chessboard, the king would have to pay more than nine *quintillion* grains (9,000,000,000,000,000,000). Worse, when all of the squares were added together, more than 18 quintillion grains were owed to the wise man—a number far greater than all the grains of rice ever grown in the history of the world. See Figure 3.1.

At this point, some versions of the legend say the king put the wise man to death for his manipulative strategy. In other versions, the king instead insisted he be relieved of the unpayable debt, and also that the wise man agree to become the king's new, highest adviser. In the

•	••	••••	•••• ••••	•••• •••• •••	•••••••• ••••••••	▦	128
256	512	1024	2048	4096	8192	16384	32768
65536	131K	262K	524K	1M	2M	4M	8M
16M	33M	67M	134M	268M	536M	1B	2B
4B	8B	17B	34B	68B	137B	274B	549B
1T	2T	4T	8T	17T	35T	70T	140T
281T	562T	1Q	2Q	4Q	9Q	18Q	36Q
72Q	144Q	288Q	576Q	1QT	2QT	4QT	9QT

Figure 3.1: The king's chessboard illustrates the extraordinary properties of exponential growth, and how quickly numbers expand beyond quantities and scales normally encountered in everyday human life.

most interesting ending, however, the king reflected carefully for a few moments . . . and then told the wise man, "Fine. I will pay you. But before you can leave, you'll have to count each of the grains I give you." Although the wise man objected, saying he'd rather be on his way, the ruler firmly insisted, arguing that the wise man's count was necessary to prove the king had paid his debt honorably and entirely. In truth, the king had simply learned well from the lesson he'd just been taught. *If it takes one second to count a single grain of rice, it would have taken the visiting wise man more than a half-trillion years to count them all, over 40 times the age of the universe itself.*

The legend of the king's chessboard illustrates how quickly enormous numbers result from repeated doubling, even on a chessboard that only has 64 squares. In the first few steps, the increases are easy to project and seem entirely reasonable. But, as with all exponential growth, once the numbers exceed the scales with which we're familiar, they escalate so suddenly that we soon lose all human frames of reference and perspective. They become incomprehensible.

———

In today's world of computing, the kinds of numbers that exceed our natural comprehension have become commonplace. Scientists, computer designers, software programmers, and even consumers encounter them every day. They explain our universe, our products, and even ourselves. They also explain artificial intelligence.

In order to express, calculate, and discuss extremely large numbers in a manageable format, we use a method called *scientific notation*. It's how mathematicians and scientists express otherwise unwieldy numbers for conversational and rounded computational purposes. Although the concept can be a bit confusing at first, a careful reading of the next two paragraphs will prove useful in later chapters.

In scientific notation, an exponent is a number that's placed to the right and slightly above another number to indicate how many times the first number should be multiplied by itself. The exponent represents the *order of magnitude,* or the *power,* by which the first number is raised. For instance, 100 can be written as 10^2, which means 10×10 and can be thought of as a one followed by two zeroes. Similarly, 1,000 is written as 10^3, which is equivalent to $10 \times 10 \times 10$. Likewise, 10^9 means $10 \times 10 \times 10 \times 10 \times 10 \times 10 \times 10 \times 10 \times 10$. So, rather than writing 1,000,000,000, we can instead write 10^9 (meaning one with nine zeroes behind it) to express one billion.[5]

Numbers other than 10 can also be raised by any order of magnitude. In the story of the king's chessboard, for example, the number of grains the king owed for the 64th square is 2^{63}, which means two multiplied by itself 63 times (from the second square of the board onward). Scientific notation can also be used to express negative numbers and decimals. Negative one million (−1,000,000) can be written as $−10^6$. And the decimal 0.0000000056 can be written as 5.6×10^{-9}. Again, this method of shorthand will become quite relevant in later chapters.

What's particularly important to recognize is that the *differences* between two numbers written in scientific notation might not seem significant at first blush, but they often are. For instance, 10^{12} might not appear as though it would be much smaller than 10^{15}, but the difference is actually astronomical. It is the difference between one trillion and one quadrillion. That difference is equivalent to *1,000 trillion*—an unimaginably large number in and of itself. In the realm of modern computing and artificial intelligence, as we'll later see, these kinds of numbers—along with the extreme magnitudes of differences *between* them—have significant meaning and implication.

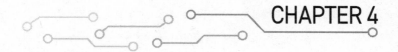

CHAPTER 4

SECRET ORIGINS OF MODERN COMPUTING

As a general rule, the most successful man in life is the man who has the best information.

> —*Benjamin Disraeli, 1804–1881*
> *Former Prime Minister, United Kingdom*

Because big numbers and the information they represent cause a haze of confusion in the human brain, we eventually required a tool to clear the air. In the early nineteenth century, a war between Spain and Mexico set in motion a strange string of events that ultimately led, more than 100 years later, to the creation of the world's first computer capable of processing amounts of information far beyond our own capacity.

After more than 300 years of occupation, Mexico's demand for separation from Spanish rule led to the outbreak of the Mexican War of Independence in 1810. Following a decade of military and insurgent fighting, the conflict finally came to an end when Mexico was granted formal sovereignty over its own land and citizens in 1821. At the time, Mexico's geographic boundaries covered a huge expanse that reached below the Yucatán Peninsula in the south and extended above what is now the American state of California in the north. To the northeast, Mexico's territory stretched all the way through present-day Texas.[1]

Despite Mexico's independence from Spain, not everyone in its widespread territories was satisfied with Mexican governance. Fourteen years after the war's end, and after many years of effort to obtain some level of its own independence from Mexico, the territory of Texas (historically referred to as "Mexican Texas") formally revolted and fought a number of intense battles against Mexican troops—*Remember the Alamo*. Texas soon prevailed and, in 1836, Mexico was reluctantly forced to recognize the independent Republic of Texas.[2] Mexican authorities were never fully accepting of Texas's sovereignty and were even more outraged when the American federal government annexed the Republic of Texas in 1845, by making it the 28th state of the expanding American union.[3]

Combined with continuing border disputes and the US government's apparent interest in obtaining even more Mexican territories, the US annexation of Texas led to the Mexican-American War in 1846. After two years of conflict, the Treaty of Guadalupe Hidalgo brought the war to an end. The treaty established the Rio Grande as the permanent border between the two countries and required Mexico to relinquish more than 55 percent of its total territory in exchange for relief from $15 million owed to the US government and another $3.25 million owed to individual US citizens.[4] Objectively, the 1848 treaty was an unequivocal loss for Mexico, and few Americans today realize the immense amount of territory relinquished to the US. The lands turned over included all of present-day California, Nevada, and Utah; most of

Arizona; approximately half of New Mexico; a quarter of Colorado; and a small section of Wyoming.[5]

So, what does Mexico's independence from Spain combined with Texas's subsequent independence from Mexico, America's resulting annexation of Texas, and the American acquisition of more than half of Mexico's territory at the end of the Mexican-American War possibly have to do with computer technology and the eventual creation of artificial intelligence? Fast-forward seven decades, all the way to the other side of the Atlantic, and the connections unfold.

On June 28, 1914, Austria's archduke Franz Ferdinand and his wife were killed by a Serbian nationalist in Sarajevo. A month later, Austria-Hungary responded to the assassinations by declaring war on Serbia. Within only a few days, Germany, Russia, France, Belgium, Britain, and Japan had all entered the fray. It was the start of World War I.

Over the next two years, the war spread throughout Europe and into Russia and East Asia. Although the US remained neutral in the war's initial years, Germany became increasingly convinced the US would eventually get involved—especially since the German navy was preparing in early 1917 to launch a submarine campaign targeting American cargo ships and Allied vessels in the North Atlantic. Knowing that such attacks would likely prompt the US to enter the war, the Germans looked for some way to discourage American involvement.

The strategy they devised was to distract the US by causing a war on its own soil. Reasoning that a North American conflict might preoccupy the Americans and dissuade them from sending any troops across the ocean, Germany looked back in history to the Mexican-American War. Even though it had ended almost 70 years earlier, the Germans knew that tensions and distrust still remained between Mexico and the US.

In an effort to take advantage of lingering animosities, the German secretary of foreign affairs, Arthur Zimmermann, sent a coded telegram on January 19, 1917, to Germany's ambassador in Mexico instructing him to offer an alliance and financial support to Mexico *if* it would agree to invade America should the US enter the war. In pertinent part, the telegram read:

> We intend to begin on the first of February unrestricted submarine warfare. We shall endeavor in spite of this to keep the United States of America neutral. In the event of this not succeeding, we make Mexico a proposal of alliance on the following basis: make war together, make peace together, generous financial support and an understanding on our part that Mexico is to reconquer the lost territory in Texas, New Mexico, and Arizona . . . Signed, Zimmermann.[6]

See Figure 4.1.

Though little is made of it in history books, Zimmermann's telegram was one of the most significant strategic missteps in military history. It not only failed, but completely backfired. Unknown to the Germans, the British had been intercepting their military signals and communications for years. When the Zimmermann telegram was sent, the Royal Navy's code-breaking operation intercepted it, deciphered it, and, a little more than a month later, turned it over to the American embassy in London.[7]

On February 26, 1917, American president Woodrow Wilson first learned of the telegram. At that point, the German submarine campaign in the North Atlantic had already begun, and American cargo ships were sinking just as the telegram foreshadowed. Although Wilson was already strategizing a military response, many Americans and members of Congress were still strongly opposed to entering the war. But the Zimmermann telegram was Wilson's ticket to change public opinion. He presented it to Congress and instructed the State Department to openly release its contents to the American media.[8]

On March 1, 1917, news of Germany's secret proposal to Mexico made newspaper headlines across America.[9] Even then, some Americans thought it was contrived propaganda—fake news, as it's now called. But two days later, Zimmermann himself acknowledged the telegram's

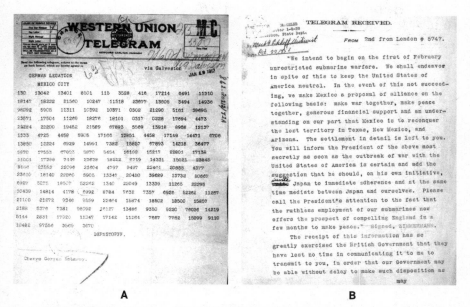

Figure 4.1: (A) The January 19, 1917, coded telegram sent from German foreign minister Arthur Zimmermann. (B) The decoded translation of the intercepted telegram, accomplished by the British Royal Navy's code-breaking operation and later shared with American intelligence.

authenticity. With the will of the people then unified and strongly behind him, Wilson confidently requested a declaration of war from the American Congress. On April 6, 1917, the US formally entered World War I.

———

In a strange but certain way, the British intercept of the Zimmermann telegram provided the catalyst for modern electrical and digital computing. After World War I, paranoia and a sense of urgency in Germany and elsewhere to protect the secrecy of information and communications was rampant. The lessons learned during the war about the value of keeping coded information secret, coupled with the growth of international commerce and the expanding use of the telegraph, created a critical need for a way that governments, militaries, and private companies could keep their communications secret and safe from their enemies and competitors.

As a solution, an enterprising and inventive German electrical engineer named Arthur Scherbius patented an affordable machine capable of encoding information with a level of complexity far beyond anything previously possible. In 1923, Scherbius set up the Cipher Machines Corporation in Berlin to manufacture his product.[10] He called it Enigma. Within a few years, the machine and a variety of foreign design variations were being used by governments and major industries throughout Europe and the wider world to encode much of the confidential and proprietary information contained in their daily telegrams.

Scherbius's Enigma required an operator to type the original letters of an outgoing message into the machine one letter at a time, just as you would on a traditional typewriter or modern keyboard. Through a series of electrical connections and interchangeable, notched mechanical wheels—each of which had 26 different letter positions—Enigma would scramble each of the input letters and then output an entirely different and unpredictable coded letter. Every time a single letter was input and scrambled, the mechanical system of the machine would itself scramble, by advancing its mechanical rotor wheels so that the next letter would be coded based on an entirely different series of machine connections. The number of different possible output variations from each keystroke was staggering, and deciphering its code was a practical impossibility.

Once completed, the entire coded message could be transmitted in its newly encrypted format, usually via Morse code. Assuming the recipient of the message at the other end of the telegram had his or her own Enigma machine, *and* that he or she knew the exact starting arrangement and positions of the internal rotors the sender had set on the originating Enigma before beginning to scramble the message, the recipient could unscramble the code through Enigma's reverse deciphering process.

The German military began using Scherbius's Enigma machine in 1925. Within just a few years, it became the primary means of encrypting messages in the German army, air force, and navy. As they adopted it to

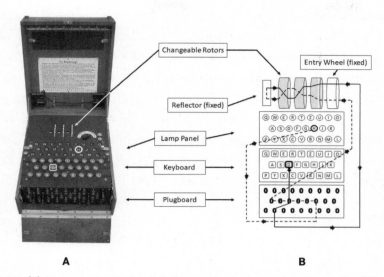

Figure 4.2: (A) German World War II Enigma with corresponding illustration (B) of the Enigma's input and output paths. When a key is struck on the Enigma's keyboard, the signal travels out through the plugboard to the entry wheel, then through the changeable rotors to the reflector, and then back again through the rotors, the entry wheel, the plugboard, and into the lamp panel. In the example shown, a keystroke of the letter *D* results in a coded output of the letter *H*. Each keystroke alters the starting positions of the rotors for the next keystroke. Because of the changeable plug and rotor configurations, a German military-grade Enigma had 159 quintillion (158,962,555,217,826,360,000) different possible pathways from a single keystroke to a coded output.

their operations, Germany's military coding experts and engineers made the basic design of Enigma even more complicated by adding additional rotors, electronic circuits, and plugs.[11] See Figure 4.2.

———

Fifteen years later, on September 1, 1939, Adolf Hitler's Nazi Germany invaded Poland. Within days, Britain and France responded by declaring war on Germany. World War II was underway . . . and Enigma was destined to play a critical role. At the outbreak of the war, the Nazis' enhanced version of the machine allowed for 159 quintillion (15.9×10^{18})

different possible settings and paths to create an encoded output for every single keystroke entered into it.[12] The number "quintillion" was so unimaginable at the time that the word wasn't even used. Instead, even mathematicians simply called the number "159 million, million, million."

To make matters worse, all of the German Enigma operators simultaneously changed their mutual configurations of the rotor wheels and other variable features on their Enigma machines every single day, at the stroke of midnight. As a practical matter, without having an Enigma machine *and* without knowing the exact daily positions of all rotors and plugs, it was impossible for any human to crack the code.

By 1940, fleets of German submarines (called U-boats, which is short for *Unterseeboot*—meaning "undersea boat") were again attacking shipping lanes throughout the North Atlantic. It was a replay of World War I. This time, though, the German submarines were coordinating their attacks by communicating in Enigma code. At the same time, the German army and air force commands were also using Enigma to coordinate infantry and armament movements and attacks throughout Europe. The airwaves were alive with German radio and Morse code signals, but even though the Allies could easily listen to the Nazi transmissions, they were of no use in their indecipherable, Enigma-encoded formats.

Cracking the daily Enigma settings was therefore the key to penetrating the German military's vast language of information. If it could be accomplished, then the Allies might be able to determine strategies and solutions to combat the Nazi threat. To tackle the Enigma problem, the British government established the Government Code and Cypher School (GC&CS) at the site of a former radio factory called Bletchley Park, 50 miles northwest of London. They gathered a small and secret team of their best code breakers, linguists, chess players, crossword enthusiasts, and mathematicians. The team included Alan Turing, a mathematician who before the war had solved a famous mathematical problem by hypothesizing what he termed a universal computing machine. Because of that and the work he was about to accomplish at Bletchley, he later became known as one of the founders of modern computing and

AI.[13] Just recently, in 2019, the Bank of England chose Turing as the face of Britain's new 50-pound bank note, announcing that, "as the father of computer science and artificial intelligence, as well as war hero, Alan Turing's contributions were far-ranging and pathbreaking . . . [he was] a giant on whose shoulders so many now stand."[14]

With the single goal of deciphering the German military's daily Enigma settings, the code breakers at Bletchley Park worked tirelessly to come up with a solution. But the problem seemed insurmountable. There were just too many variables to solve—too many available settings and too many combinations that, in total, allowed for 159 quintillion possibilities. It was an unmanageable number. Even if they had a team of thousands, and even if each person worked around the clock attempting to go through every potential setting and output Enigma could create, it would still take thousands of years to process all the possibilities. They didn't have thousands of years. They barely had thousands of minutes. To be precise, they only had 1,440 minutes—24 hours—between each time the Germans changed to a new group of Enigma settings, making any progress the code breakers might have made in the interim completely irrelevant.

Eventually, Turing came up with two breakthroughs to solve what seemed the unsolvable. First, he realized that certain German messages, sent at particular times of the day and for particular reasons, would always have specific repetitive and patterned formats. Daily weather reports that were strictly reported at 6:00 AM from the North Atlantic German U-boats were the best examples. Turing and his team knew that every morning those reports would follow a consistent format: Date, time, wind speed, atmospheric pressure, and temperature. In an attempt to shortcut the code-breaking process, Turing used the recurring patterns of information in the weather reports as the basis for a crib, a short section of a coded message that he anticipated would include certain words. Essentially, he was making educated guesses about what the crib *should* say, and then working backward to see if the code could be deduced. It was an innovative idea at the time that would later become an elementary root of modern computer analysis—to methodically search for a specific piece or pattern of information suspected to exist somewhere within a greater field of information.

Even with that advantage, Turing soon realized that it would still take far too many man-hours to sift, one by one, through all of the possible combinations of Enigma settings, regardless of how many people were available to work on the task. He therefore coupled the crib approach with his second breakthrough, built upon earlier work by Polish collaborators—the design of an immensely complex, electromechanical machine capable of sorting quickly and methodically through massive numbers of possibilities.

Turing's new machine generated intermittent electrical circuits that ran through every possible Enigma setting—through reverse deciphering—in search of the one complete circuit that tracked back to the original letters of a particular crib. If the machine could find a combination of settings that accurately replicated the entirety of just one of the cribs taken from the day's German radio traffic, then those settings would also decrypt every other German message sent the same day. After complicated and painstaking modifications to his original design, Turing's machine eventually worked just as envisioned . . . and successfully deciphered German radio transmissions on a regular and reliable basis.

Turing called his machine the Bombe after the earlier, manually operated Polish Bomba. Turing's was an extravagantly complicated machine for the time, built of more than 12 miles of electrical wiring and more than 97,000 mechanical parts.[15] The Bombe's significance for modern computing is that it was the first machine to capably sort through tremendous amounts of variable information to find a solution that a human couldn't possibly accomplish on his or her own. That same basic purpose is essentially what today's computers do. And although modern computers are infinitely more refined than the Bombe, they are, at the core of their function, arguably no different.

By the end of the war, there were over 200 Bombe machines in operation throughout England, all working around the clock to decipher

German radio traffic every day.[16] Amazingly, the Germans never learned of Bletchley Park, nor did they ever discover that their radio communications were being deciphered on a continual basis. To avoid revealing that Enigma had been solved, the British and Allied forces maintained the secrecy of their code-breaking ability as an absolute, and never even acted on most of the intelligence they gained from the deciphered German messages. Even so, it's estimated that the information the Allies acquired through Turing's Bombe, along with other work accomplished at Bletchley,[17] shortened the war by at least two years and saved millions of lives.[18]

——

After the war, Alan Turing continued to be a driving force in the fields of theoretical and practical computing. Other than his work on Enigma, he's best known outside of academic and research circles for a concept known as the Turing Test, or the *Imitation Game.* He introduced the concept in a 1950 paper titled "Computing Machinery and Intelligence."[19] The first sentence of the paper is: "I propose to consider the question, 'Can machines think?'"[20]

The paper describes a test to measure a machine's ability to exhibit intelligent behavior indistinguishable from that of a human. In Turing's description of the test, a human evaluator acts as the judge of a conversation that includes himself, another human being, and a machine designed to generate humanlike responses. All three of the participants are kept physically and visually separate from each other, and the human evaluator isn't told which of the other two participants is the human and which is the machine. The three of them then participate in a text-only conversation to determine if the human evaluator can reliably distinguish between the machine and the other human based on the content of their replies and interactions. The results are dependent not on the machine's ability to provide correct or accurate answers, but on how closely the answers and language used to express them resemble those that a human would naturally give. If the machine can accomplish an indiscernible *imitation*, then it passes the test.[21] Since Turing first introduced the concept of the imitation game, it has continued to be highly relevant for research, development, and application and ultimately became an important foundational element in the philosophy and technology of artificial intelligence.

Following World War II, digital computer technology continued to progress at a steady pace. Mechanical and electrical designs and production, particularly the development of transistors and circuits that we'll discuss in the next chapter,[22] became more efficient and sophisticated. Most importantly, computer languages were introduced that gradually made computers more adaptable to general use. It was the development of those languages that set the stage for all computer languages developed afterward. And it's only through those later computer languages that we can now program and instruct present-day digital computers to perform all of their functions, including artificial intelligence programs and machine learning applications. As we're about to see, computer languages, whether basic or complex, are essentially no different than our own.

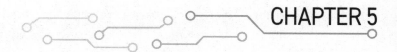

CHAPTER 5

UNIFYING THE LANGUAGES OF
MEN AND MACHINES

Mathematical science shows what is. It is the language of unseen relations between things. But to use and apply that language, we must be able fully to appreciate, to feel, to seize the unseen, the unconscious.

—Ada Lovelace, 1815–1852
English Mathematician and Computing Theorist

The world is a diverse place, with more than 7,000 distinctly different spoken languages. Of those, roughly 4,000 have corresponding written forms.[1] Despite those numbers, more than 88 percent of the world's population speaks at least one of the ten most common languages as their native tongue. In descending order of use, they are English, Chinese/Mandarin, Hindi, Spanish, French, Arabic,

Bengali, Russian, Portuguese, and Indonesian.[2] Almost two-thirds of the world's population is bilingual or multilingual, and more than two-thirds of the entire population speaks one of the three most common languages—English, Chinese/Mandarin, or Hindi—as either their sole or secondary language.[3]

Traditionally, our human languages (which are commonly referred to as *natural* languages) were used only to exchange information directly from one person to another when both shared a common language. When they didn't, the best solution was to use an interpreter—someone who knew both languages—to facilitate the conversation. Historically, language interpreters obviously needed to be humans. But, with the advent of digital technology, that's no longer the case. Natural language processing (NLP) programs and machine learning applications, which we'll discuss more in Chapter 9, are now readily available that can interpret human language (whether spoken or written), determine its meaning, execute commands, and even translate into other natural human languages in real time. As just one example, Google Translate offers varying degrees of support for more than 100 languages, with translation of typewritten submissions available for all of them, and interactive, immediate speech-to-speech translation available for most.[4]

Although computers are becoming increasingly able to process our languages, they're only able to do so after they've been initially programmed by a human who understands computer speak. In today's world, where information of every type imaginable is being created, collected, stored, analyzed, and exchanged by computers and other digital devices, computer languages—which are the means through which we empower and instruct computers to understand and manipulate the data we provide them—are becoming arguably as significant to us as any of our own natural languages.

At the core of all languages are two building blocks: the first is the complete vocabulary of words available in the language (including all of their individual definitions), and the second is the set of rules for putting those words into valid combinations to create sentences that can express more complex and refined meanings. Through the rules that allow certain words to be combined with others, the finite number of words in any one language or in any one person's vocabulary can be endlessly combined and strung together to create limitless degrees of specificity and detail. Said another way, words that have a single meaning when standing alone can be combined to create an infinite possible number of differently nuanced descriptions, narratives, commands, requests, or inquiries.

The meanings of individual and combined words are referred to as the *semantics* of a language, whereas the system of rules that govern how those words can be combined and structured into grammatically correct sentences is called *syntax*. All languages have semantics and syntax. And to be functional or fluent in any language, you must have a reasonable command of both.

————

Computer programming languages are no different. Often referred to as *constrained* languages, they were first created for very limited, specialized purposes. The first commercially available high-level computer programming language was published in 1957 by IBM. It was called FORTRAN (FORmula TRANslation), and it gave computer programmers a way to express scientific and mathematical computations similar to the mathematical formulas themselves.[5] In 1964, two professors from Dartmouth ran the first program written in a new, simpler language they created, called BASIC (Beginners All-Purpose Symbolic Instruction Code). It was much easier to use than FORTRAN and gave people other than computer scientists the ability to work more easily within its language and programming rules. It was also the first compiler, or translator language, that early computer manufacturers included as part of their systems.[6] Since then, thousands of additional computer languages have been created, more are developed every day, and their purposes are becoming increasingly more expansive—particularly as a result of current AI and machine learning technologies. Like natural human languages, however, there's a comparatively small number that are most often and widely used.[7]

Science fiction aside, no computer programming language will ever be any person's native or natural language. But no one can dispute that in today's computer-driven age of information, common computer languages like Java, JavaScript, C, C++, Python, Swift, and PHP are extraordinarily useful and powerful skills to possess—so much so that they're now becoming accepted in various places as the equivalents of true secondary languages. In lower and upper school systems around the world, there is a growing movement to give computer programming skills and fluencies academic credit equal to traditional foreign languages. Many educational systems now even allow proficiency in a computer language to satisfy foreign language curriculum and graduation requirements. For our purposes here, it's not important whether computer languages are considered academically or culturally acceptable *replacements* for some other secondary natural language. It's enough only to recognize that computer programming languages are, for all intents and purposes, legitimate languages unto themselves—and undeniably useful *as* secondary languages going into the future.

Just like natural human languages, computer languages—including those used to write artificial intelligence and machine learning algorithms—have their own vocabularies and their own sets of grammar and construction rules. They have their own semantics and their own syntax. Unlike human languages, however, they don't have any general tolerance for ambiguity, error, subjective interpretation, or unspecified rules or expectations. To understand why that is, in fact why that *must* be, along with the basics of how these languages are constructed and how they operate, it's necessary to first understand the fundamentals of how digital computers work.

Many of us know or have at least heard that everything a computer does, and everything a computer is capable of doing, ultimately comes down to ones and zeroes. But what does that actually mean? You'll recall from the previous chapter that different numerical systems were developed and used by different cultures during the course of human history. By the late Middle Ages, most of the planet's population had settled upon Arabic numerals, which is the same base-10 decimal system we use today. To refresh, base 10 is a positional system with ten distinct symbols—including nine numerical characters and a zero placeholder (1 through 9, and 0). When counting in base 10, we proceed through the singles digits by using the numerals 1 through 9, and then, when it's necessary to represent a tenth unit, we place a 0 symbol in the singles' digit column and add a 1 numeral (positionally) in the tens' column to its left, to represent 10. Then, we repeat the pattern of counting upward in the singles' digit column (11 through 19) until we again place a 0 symbol in the singles' column and replace the 1 with a 2 in the tens' column, to represent 20 (which, literally, stands for two groups of ten units).

Computers, however, don't operate on a base-10 system. They utilize a base-2 system, also referred to as binary. Conceptually, base 2 is not really that different than base 10. Visually, though, it looks very different and counting can be confusing. That's because base 2 only has two distinct symbols (1 and 0) to work with. Even so, with a bit of careful explanation, the way computers count and calculate numbers in binary isn't that difficult to understand.

Rather than counting upward from 0 to 9 before adding another digits column to the left in order to represent the next exponential category of units, binary only counts from 0 to 1 before requiring an additional column. Each new digits column therefore represents an increasing exponential power of 2, rather than 10.

Although this sounds complicated, it's actually quite easy. Just as in base 10, in binary a quantity of zero units is represented as 0 and a quantity of one unit is represented as 1. However, because binary doesn't have any additional symbols available to represent additional individual units (i.e., there are no 2 through 9 numerals), two units cannot be represented as "2." Instead, two units must be represented in binary as "10" (meaning one value of 2, plus zero values of 1). "Three" units are represented as 11 (meaning one value of 2, plus one value of 1). Likewise, "four" units are represented as 100 (meaning one value of 4, plus zero values of two, plus zero values of 1). The process continues expanding to the left as number values continue to increase—but still only with 1 and 0 symbols.

So, why do computers use the base-2 binary system? Why not base 10, or even some other system? The reason is simply because of the basics of mechanical and electrical componentry. And those basics go to the very heart of computer technology itself. At their core, modern digital computers operate through a system of micro transistors, which are incredibly small, artificially fabricated components that only have one functional purpose. Through the impulse of an extremely small electrical charge, they can be switched to an "on" position. When they don't have an electrical charge, they remain in an "off" position. Depending upon their position at the exact instant of any computer calculation or process, each transistor communicates a single binary fragment of information to the computer's processing unit. If an individual transistor is charged, or "on," the computer interprets the signal as representing an affirmative 1 symbol—an indication of "yes." If the transistor isn't charged, or is "off," the computer interprets the information from that particular transistor to be a 0 symbol—an indication of "no."

When a computer looks at a group of transistors arranged in a linear format, it interprets the combined "on" and "off" positions of all of the group's transistors at a simultaneous instant in time to interpret a combined total sum that the transistors, collectively, represent (as a combination of 1s and 0s, in binary format). To make this a bit clearer, in computer terminology, each single transistor is referred to as a *bit*, which stands for *binary digit*. Early computers were capable of recognizing and interpreting eight transistors grouped, in a line, together. That grouping of eight transistors is known as one *byte* (8 bits equal 1 byte).

In the base-2 binary system, eight transistors arranged in a linear format (each of which can individually be turned either "on" or "off" at a simultaneous moment) collectively represent eight digital columns,

Converting from Base-10 to Base-2								
Base-10 Value	Base-2 Representation, in a Binary 8-Bit Format							
	128 +	64 +	32 +	16 +	08 +	04 +	02 +	01
A. 0	○	○	○	○	○	○	○	○
1	○	○	○	○	○	○	○	●
2	○	○	○	○	○	○	●	○
3	○	○	○	○	○	○	●	●
4	○	○	○	○	○	●	○	○
B. 7	○	○	○	○	○	●	●	●
8	○	○	○	○	●	○	○	○
10	○	○	○	○	●	○	●	○
18	○	○	○	●	○	○	●	○
22	○	○	○	●	○	●	●	○
36	○	○	●	○	○	●	○	○
42	○	○	●	○	●	○	●	○
C. 73	○	●	○	○	●	○	○	●
116	○	●	●	●	○	●	○	○
154	●	○	○	●	●	○	●	○
203	●	●	○	○	●	○	●	●
D. 255	●	●	●	●	●	●	●	●

Figure 5.1: Converting from base 10 to base 2 binary counting in a digital 8-bit (1-byte) linear format. ○ symbols represent uncharged "off" transistors. ● symbols represent charged "on" transistors.

or exponential powers of 2 in binary counting. When their individual values are added together, they can represent any number from 0 through 255. For instance, if all eight transistors are simultaneously in the uncharged "off" position, the computer interprets them as having a total combined value of zero. See Figure 5.1, line A. On the other hand, if the first, second, and third transistors (from right to left) are charged at a particular instant, then the representative value of the overall byte is seven (i.e., 0 + 0 + 0 + 0 + 0 + 4 + 2 +1). See Figure 5.1, line B. Alternatively, if the first, fourth, and seventh transistors (again, from right to left) are charged, the representative value of the byte is 73 (i.e., 0 + 64 + 0 + 0 + 8 + 0 + 0 + 1). See Figure 5.1, line C. And, as a last example, if all eight are simultaneously in the charged "on" position, then their combined, full value is 255 (i.e., 128 + 64 + 32 + 16 + 8 + 4 + 2 + 1). See Figure 5.1, line D. (*If you don't have any previous experience with the binary system or the way transistors function in a digital computer, this would be a good time to pause and carefully review Figure 5.1, just to absorb the basic concepts. You don't need to master binary counting, but it's very helpful to understand the principles because, as we're about to see, the binary transistor system is the basis upon which every function of every digital computer depends.)*

Since binary is simply a system of counting, you might be wondering how any information *other* than numbers can be processed or interpreted by a computer. That's an insightful question. Another way of asking is, "How does a computer *translate* from numbers to language, or from numbers to any nonnumerical information (like audio sounds or visual images)?" The easiest way to explain the process is to describe how computers convert from numbers to the letters of our natural language—which, in essence, is what happens each time we type a word onto our keyboard and see it immediately appear on our screen. It's also what our computers do each time they interpret an incoming email and display it to us in English, Spanish, or whatever other natural language we speak.

When we type or spell anything on a computer or other digital device, what we're really doing is instructing the computer to interpret a digital keystroke—through its internal binary operations—by following a standard programming code called ASCII (pronounced as-kee), which stands for the *American Standard Code for Information Interchange*. ASCII is a text-encoding format for computers that was first created in 1963, and then widely standardized when IBM used it on the public release of its first PC in 1981. ASCII converts letters of text into numbers so they can be read, stored, and processed by a computer. In ASCII's encoding format, each letter of the alphabet is assigned a corresponding base-10 numeric value, and then that value is translated to its base-2 format so it can be represented in an eight-transistor byte.

Because there are 256 different possible values that can be represented in a base-2 format by any eight-transistor byte (including 0 through

255, as explained above), ASCII has 256 different numeric values that are utilized to represent different letters, characters, and other symbols. That's enough to assign an individual value to each of the 26 letters of our English alphabet, in both lower- and uppercase (therefore requiring 52 numeric assignments), with 204 additional values left over—which are assigned to other common characters, symbols, and punctuation marks.[8] For instance, the ASCII value assigned to uppercase S is 83. So, whenever you type an S on your keypad, deep within your computer your keystroke is translated to the base-2 representation of 83 and then back again to a capital S symbol appearing on your screen.

This process of interpretation, conversion, and execution is all accomplished in an imperceptibly small instant of time, and it's all facilitated by a single group of micro transistors that, when you strike the S key on your keyboard, are instantly switched to individual "on" and "off" positions on a single byte (8-bit) configuration that, in binary counting, represent 83 (represented as 0-1-0-1-0-0-1-1 in an 8-bit transistor configuration). Similarly, the ASCII value for an exclamation mark is 33, which is represented as 0-0-1-0-0-0-0-1 in an 8-bit binary format. That same process occurs for every other numeric interpretation that computers make and that result in any execution they consequently take, whether it is the next letter or character in an email you're typing, a single pixel display of color flashed for a fraction of an instant as part of a video stream, or the next step in some other programmed calculation.

As you might be imagining, it became clear very early in the development of digital computers (the evolution of which we'll discuss in the next chapter) that single-byte groupings of transistors—with possible combined number values only as high as 255—just weren't going to be sufficient. For instance, even if a computer adds the separate values of four active bytes together, the highest number it could interpret from them would be 1,020 (255 + 255 + 255 + 255). The solution to this capacity problem was to design new computers capable of recognizing two bytes of transistors *linked together* (as 16 transistors, rather than 8). Instead of referencing only 8 digital spots of available binary values as an 8-bit processor does, a "16-bit" processor can instead recognize 16 digital spots of available values. This increased the informational and computational capacities of computers *exponentially* from 2^8 to 2^{16}. The effect of the change was substantial. It increased the possible number of values that could be represented in any one instant by an aligned group of transistors from 255 (as in one-byte of 8 bits) to 65,535 (by utilizing all 16 bits aligned in a two-byte group of transistors). And the exponential increase in a modern 64-bit computer (which references up to 64 individual transistors, or eight total bytes) is even more pronounced. It can potentially reference up to *18 quintillion* distinct numbers and, hence, 18 quintillion distinct pieces of information. Those distinct pieces of information that a 64-bit computer can potentially read at any one instant are analogous to the 18 quintillion individual grains of rice that were owed to the wise man in the Chapter 3 story about the king's chessboard. The exponential process of growth is identical in both cases—64 transistors and 64 squares on the chessboard, with each one of them being a spot that represents two times the numerical value of the transistor or square before it.

That, however, is only a part of the equation. It's important to realize that *each* 64-bit compilation of information that a computer can reference (through the combined positional "on" and "off" states of all 64 transistors in its 8-byte system) only represents one piece of every calculation that a computer is capable of making. And, as we indicated back in Chapter 3, today's computers can make billions of calculations every second.

———

As a practical matter, all of this means that in order to utilize the physical and electrical potential of its technology, a computer must have billions of transistors operating inside of it. But how is that physically possible? The answer lies in the microscopically small nanoengineering capabilities behind today's computer-chip production methods. Modern transistors are made of silicon, which, behind oxygen, is our planet's second most abundant element.[9] As a result of current nanoscale design and manufacturing processes, today's common production transistors are incomprehensibly small—less than 14 *nanometers* wide. A millimeter (10^{-3}) is one-thousandth of a meter, and you might recall from Chapter 3 that 25.4 of them fit in the length of an inch. By comparison, a nanometer (10^{-9}) is one-billionth of a meter . . . and 1,500 of them fit within the width of a single strand of human hair. With that kind of current technology, more than 19 billion transistors can be engineered and produced to fit on a single integrated circuit board the size of a fingernail. What's more, manufacturers are now transitioning to the production of 10-nanometer transistors, and are planning for 7- and even 5-nanometer chips within the next few years.[10]

All of these advances to fit ever more transistors into microscopic chip space on tiny circuit boards continue to increase the ultimate computing powers and speeds of computers. For many years, computing power had been doubling approximately every 18 to 24 months. That rate of predictable technological growth is referred to as Moore's Law—since it was first demonstrated by Gordon Moore, the cofounder of Fairchild Semiconductor and Intel Corporation, back in 1965.[11] Not a true physical law, but only a prediction of performance capabilities based on the rate of technology growth and more advanced production processes, Moore's prediction that microchip performance capabilities would continue to double on this regular, recurring timeline has essentially proved true ever since. It has been used for decades by the semiconductor industry to guide long-term planning and to set research and development targets. In fact, the rate of advance in most digital electronic technologies has been consistent with Moore's prediction as it proved equally applicable to various other componentry as well—including memory capacities, microprocessor speeds, and even the number and diminishing sizes of pixels in digital video devices. And while some argue Moore's law may be nearing its end, or at least slowing down, to this point the emergence of nanomanufacturing techniques have kept Moore's constants relatively accurate.[12]

———

Considering the fundamental, transistor-oriented operations of computers, the purposes of computer programming languages are easy to conceptually understand. A computer programming language is nothing more than a vocabulary and set of grammatical rules that programmers use for writing software, scripts, and other sets of instructions that computers follow to perform the specific task the programmer or end user wants to accomplish.

The most expedient programming languages, called *low-level*, are those written in the fundamental binary code language of the computer itself. Computers can process and follow low-level languages and programs directly and immediately, without any need for the computer to first engage a translator or interpreter program to change the programmer's source code to binary. While low-level programming languages therefore have the advantage of speed and efficiency when executed, they are exceedingly tedious and complex for humans to write.

High-level computer programming languages, on the other hand—such as Python, Java, C++, and C—are much easier to write, but they rely upon other interpreter programs (called compilers) to convert their high-level code into the machine's underlying binary code. They're therefore a bit slower for the machine to execute, but they're much more commonly used. There are also programming languages that contain both binary and high-level vocabularies, and they're accordingly referred to as mid-level languages. They have the incumbent advantages of fast computer executions but without the complexities of complicated writing requirements.

Since the earliest programming languages were developed for the first digital computing tasks, many additional computer languages have been developed. But, again, only a comparatively small number of them are most commonly used, and each of them has different characteristics, vocabularies, and grammatical construction rules that make them more or less appropriate for different types of intended uses. Some are more appropriate for analyzing large data sets, some are more appropriate for scientific or math-based calculations, and some are more appropriate for communication purposes, game play, video rendering, etc.

Regardless, just like natural human languages, all computer programming languages have two essential elements: semantics and syntax. They each have unique sets of keywords and special syntactical rules that both the programmer and computer must follow, and that allow the computer to organize and run the program's instructions within the language's applicable parameters. What sets computer languages apart, however, is that they provide a means of unifying the different natural languages people may speak into a common method of communicating— not only between themselves, but also with independent or mutually shared computers or computer systems. They are, arguably, a common and multipurpose translator for all humankind.

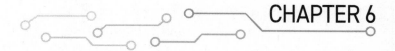

CHAPTER 6

CONSCIOUSNESS AND OTHER BRAIN-TO-COMPUTER COMPARISONS

It's ridiculous to live 100 years and only be able to remember 30 million bytes. You know, less than a compact disc. The human condition is really becoming more obsolete every minute.

—Marvin Minsky, 1927–2016
Cofounder of MIT AI Laboratory

Through a combination of silicon, plastics, metals, electrical connections, and code, the goal of artificial intelligence is to simulate the intellectual capabilities of the human brain. A biological command and cognition center that took billions of years to evolve, the brain operates in ways that are still far beyond our complete understanding. Yet, with each passing day, AI and machine learning applications replicate more and more of the natural processes and evaluative skills of which humans are capable. In an ever-expanding list of narrowly focused and single-domain tasks, computers can already learn and perform much more efficiently than us. From the internet search engines that find the information we need, to the home security systems that distinguish between our pets and unwanted intruders, to the traffic apps that evaluate road conditions and route us to our destinations most quickly, *narrow* AI that accomplishes a single type of task is, literally, already in use all around us.

While the scopes of the tasks that narrow AI programs can accomplish are broadening, computers are still not anywhere close to accomplishing the *general intelligence*, multitask potential and performance parameters of which humans are capable. General artificial intelligence, as we'll discuss in Chapter 9, is still far off, if ever. And, with the current technological means we have of accomplishing AI, it's difficult to objectively foresee that artificially intelligent computers will ever be capable of generating their own independent or self-motivated intentions—at least not without an additional breakthrough not currently visible on the horizon. Also, as far as we now know, consciousness is a phenomenon unique to biology. The magic of consciousness, the existential state that so profoundly distinguishes biological life from technology, is still an unknown dimension away.

———

The human brain is an intricate wonder of evolution. It's the most elaborate and impressive structure we know of in the universe. Our brains control and expertly manage an absurd number of responsibilities as we multitask our way through the complexities of our entire existence. From before we're born until the last breath we take, our brains constantly regulate our heartbeat, breathing, temperature, chemical balances, physical equilibrium, sensory perceptions, and all of our other life-sustaining needs. Most of those things occur unconsciously, without us having to devote any thought or intentional focus on them. And that's fortunate, because the massive volume of information that our brains automatically assess and manage every second of our lives would be impossible for us to attentively track, let alone actively control.

While all of that is going on, our brains also enable us to do the things that set us apart from every other animal on the planet. As humans, we analyze the information around us in complex, comparative, and emotional ways. We can calculate numbers and theories, learn languages, compose music, write poetry, and create art. We develop our own unique preferences and partialities. We learn to love and appreciate other people and places. We can feel love when it's returned to us, just as we can feel unimportant, slighted, or disrespected. We can castigate and empathize, feel sorrow, embarrassment, and anger. We hold grudges, but we can forgive. We ponder philosophical thoughts, make moral judgments, and develop our own sense of right and wrong. We can gather and amass information, learn from our personal and collective experiences, innovate solutions to our problems, create cultural norms, and establish social and legal cornerstones of societies. We also enact rules and laws that reflect our expectations—not only of others, but of ourselves.

Over the last few hundred years, we've learned a lot about how the biological mind works. But the majority of what we know has been discovered only in the last few decades. Cognitive and neuroscientists say we've learned more about the actual physiology of the brain in the last ten years alone than we had known in all of our prior history.[1] Even so, much remains a complete mystery.

The greatest mystery of all is consciousness itself. We have an intuitive sense of what consciousness is, but there is no universally agreed or scientifically definitive description of it, let alone an understanding of how it arises. Generally, we tend to say that consciousness is an awareness of oneself and of one's independent existence, separate and uniquely apart from others and from the physical world around us. Said differently, we think of consciousness as an awakened, existential awareness of our own existence, of our unique and individual place in the world.

Some scientists now define consciousness even more simply, as just *the ability to have a subjective experience.*

In any event, we don't know of any level of consciousness that occurs outside of animal life. As humans, we intuitively develop, early in life, an understanding that consciousness comes only with heartbeats and blood. Tellingly, it's a rare child that doesn't impute some sense of fictional consciousness into a favorite stuffed animal. On the other hand, it would be unusual to see a child talking to a nondescript, round, brown piece of plastic. But, if other pieces of plastic are pushed into it, ones that look like eyes and ears for instance, then consciousness suddenly becomes something a child will eagerly assume. This was the original design genius behind the 1949 Mr. Potato Head toy[2]—if you give a child the ability to construct something that looks biologically alive and capable of sensing input in an animal-like way, then you've also given that child the ability to create imaginary consciousness!

It's also interesting that we humans seem to recognize at a young age that other animals have lesser levels of consciousness than we do, and also that the levels of their consciousness tend to relate in some way to the complexity of their species—or the sophistication of their intelligence. For instance, we don't really suspect that either a sea coral or a clam has any real consciousness of thought, although they are animals, but we do wonder a bit about what our goldfish might be thinking, just as we uncomfortably fear what a spider might be intending. We understandably believe that our family dog is completely aware of his or her individuality apart from us, but most of us also sense that a basset hound's overall level of awareness doesn't quite match our own.

Despite our intuitive sense of what consciousness is, we still have no idea how it actually arises, at any level, and perhaps we never will. To this day, there's no consensus—even within any one branch of our sciences—exactly how consciousness emerges from the pure physiology that takes place in any biological brain.[3] Somehow, consciousness seems to just happen, to just emanate from the chemical and electrical processes taking place within brain tissue. Many scientists, neurologists included, acknowledge that the phenomenon of consciousness might simply be outside the bounds of objective science, and its true origin beyond our ability to ever definitively understand.[4] One thing of which we are certain, however, is that the totality of what we believe consciousness is arises only, at least in our world, through biology. As far as we know, and as far as anything we've yet to discover, it emanates only from living material and not from anything artificial or mechanically constructed.

Moreover, until recently, our experiences and traditional ways of thinking told us that consciousness was a prerequisite for intelligence, and that the latter could only arise from the former. We never knew of anything intelligent that wasn't also conscious. Those two phenomena seemed always, intuitively and actually, to go together.

It's therefore understandable that science fiction seems to always put the two together. In most fictional equations, intelligence always equals consciousness. It's reasonable, then, for people to feel uneasy about, and suspicious of, machines that can learn, especially when they can do so on their own and without any continuing programming or specific oversight from us. But, in the new world ahead of us, we have to put our unease aside, right along with our old notions that intelligence always requires or results in consciousness. It doesn't.

That said, it's important to understand the very real fact that machines can exhibit some characteristics we've normally attributed only to biological consciousness. Michio Kaku—a renowned author, speaker, and professor of theoretical physics at the City University of New York—has a theory of consciousness that illustrates this point quite well. In his book *The Future of the Mind*,[5] Kaku suggests that we should think of consciousness as a progressive collection of ever-increasing factors that animals use to determine and measure their place in both space and time, and in order to accomplish certain goals.

Kaku proposes that there are three fundamentally distinguishable levels of consciousness. Level one is a creature's singular ability to understand its position in space. In other words, it's the ability to be aware of one's own spatial existence with respect to the existence of others. This is the most minimal, basic level of consciousness, and it emanates from the oldest, most prehistoric part of the brain—the hindbrain or reptilian brain. A lizard, for example, can be said to have level-one consciousness because it is aware of its own space in relation to the space of the other animals upon which it preys. The lizard needs to be aware of other animals as potential threats, for mating opportunities, and as potential food sources. Essentially, it must be skilled at those three social evaluations of the world for its survival. It needs to flee when the spatial circumstances of its surroundings indicate it's under threat, and it needs to engage if it senses it has an opportunity to mate or eat. Beyond that, it's doubtful a lizard *thinks* of much else.

Level-two consciousness is an animal's ability to understand its position with respect to others—not only contrary to them, but also in concerted accord *with* them. This level of consciousness flows from the later-evolved center regions of the brain, the cerebellum, and involves emotions and an awareness of social hierarchy, protocol, deference, respect, and even courtesies. Kaku describes this level of consciousness as the monkey brain—the ability to understand and abide by social hierarchy and order within groups and communities of animals. In humans, this level of consciousness develops during the early stages of our socialization, when young children learn from their parents and others to abide by the rules of their homes and communities, to act socially responsible, and to show respect and tolerance for others.

Level-three consciousness is the ability to not only understand our social place in space with respect to the places of others, but also understand our place in time—to have an understanding of both yesterday and tomorrow. In Kaku's view, the level-three ability to reflect, consider, plan, and anticipate the future involves unique attributes of consciousness that only humans possess. He asserts, and most others would agree, that this level results from the part of the brain that most recently evolved, which is the outer and forwardmost part that sits right behind our foreheads—the prefrontal region of the neocortex. As we discussed in Chapter 2, this is where mankind's higher thinking resides, including such skills as theorizing and strategizing.

———

With a bit of extrapolation and imagination, Kaku's theory of consciousness can be helpful in keeping computers in a realistic perspective, even when programmed with artificial intelligence or machine learning capabilities. Although the theory doesn't address exactly how, physiologically, consciousness occurs, it does allow us to measure the consciousness factors that any animal, or even any *thing,* might have.

In Kaku's view, even a thermostat can be described as having one minimal *unit* of consciousness since it is able to sense the temperature around it.[6] We can loosely say, for instance, that a thermostat is aware of temperature. That doesn't mean of course that Kaku or anyone else believes the thermostat has acquired consciousness, not even level one. But we can say that a thermostat has a single, minimal element of the kind of factors we generally associate with consciousness.

Similarly, Kaku contends that flowers, which likewise fall short of level-one consciousness, can nonetheless be said to have a few additional perceptive elements beyond that of a thermostat.[7] In addition to temperature, a flower can also sense humidity, soil quality, and the angle of sunlight. The Venus flytrap, a carnivorous plant, takes things even a step further than a flower. Beyond those things that any plant can sense, the flytrap is also able to detect the presence of an insect or spider on the blades of its leaves. When it does, it folds in on itself to capture and then digest the prey. Its mechanisms are so highly specialized that it can even distinguish between living prey and nonliving stimuli, such as falling raindrops. Yet, *Little Shop of Horrors* aside, most of us wouldn't think for an instant that a Venus flytrap has anything that even remotely approaches an animal's level of overall consciousness. And we'd be right. Nonetheless, the flytrap does have some elements of awareness that are, at least generally speaking, foundational components of consciousness.

Now, with artificial intelligence and machine learning, for the first time in our history we're confronted with *machines* that exhibit intelligent characteristics and abilities that we've previously associated only with *animal* life and consciousness. Current AI programs can process information and learn to accomplish tasks in many of the same ways animals and even humans do, and they can also learn from each new analysis or accomplishment to perform even better the next time—*we'll discuss the technology and specific processes of AI at length in later chapters.* For many of us, that's troubling. But, it shouldn't be. In the past, when we wanted to intelligently design a building, or build a car, or plan an investment strategy, it took a human to do so . . . and consciousness therefore came along, naturally, with the only intelligence capable of accomplishing the task. But, with machines now capable of completing those and many other intelligent goals without us, consciousness is no longer a necessary element. Just because computers can be programmed to accomplish such tasks on their own, and even learn while doing so, it doesn't mean that they'll one day just spontaneously develop consciousness. In this new world of ours, intelligence and consciousness are *not* interdependent.

———

Turning from the ethereal concept of consciousness to more tangible comparisons between us and our digital machines, it's not uncommon to hear the human brain referred to as a computer. While that's an apt analogy for conversation and rough comparison, it's only accurate in the simplest sense. Our brains make calculations, and so do computers. Our brains collect, process, and store information, as do computers. And the processes our brains employ can loosely be compared to the design and workings of modern computers, particularly those that are being developed with artificial intelligence technologies. But the comparisons only go so far, for our brains are much, much more than three-dimensional calculating and solution devices. And even the most advanced computers of today are far less capable, complex, and wondrous than the human brain.

The physical, chemical, and electrical transactions that take place in our brains as we process information and learn are now relatively well understood. We know how information from the outside world makes its way into and through the tissue of our brains. We know which sections of our brains are triggered and involved in the processing of certain types of information, and we know which sections formulate which types of intellectual activity and thought. We also know, basically, how those sections interact with one another. We have, as a general matter at least, a pretty clear sketch of the localized activity taking place as most of our thought processes occur.

The human brain weighs about three pounds and has more than 86 billion brain cells, called neurons.[8] Each of those neurons is shaped somewhat like an expansive and intricate three-dimensional tree that has thousands of tiny, branched tips called synapses. Through those synapses, neurons can connect and transmit signals of informational significance to thousands of other neurons that surround each of them. In an extremely simple and comparative sense, biological neurons are somewhat like the manufactured computer transistors we described in the last chapter. Just as transistors function in either an "off" or "on" position, neurons are always in either an "off" (not firing) or "on" (firing) state.

That's where the similarity ends, however, because neurons are much more complex than two-mode transistors. Neurons that get sufficiently stimulated by electrical and chemical input in short enough time periods (measured in nanoseconds) turn on and fire, but they do so more like a rapidly repeating burst of jackhammer punches than like the toggled "yes" or "no" action of a manufactured transistor. In other words, rather than a simple "on" or "off" signal, the firing ("on") action of a neuron communicates a relative *level* of excitement. And that level can be even further excited by the responsive, stimulated firing of other neurons around it—if they likewise recognize and react to a stimulus as meaningful. The overall level of excitement shared by a group of neurons confirms or establishes the total significance of a stimulus. To an impressive extent, we can physically and mathematically model this network behavior through hardware and programming software—there will be much more about how computer engineers and programmers do so in Chapter 9.

Structurally, the brain is made up of hundreds of different localized neural areas where diverse functions are concentrated. Those distinct neural areas are grouped into dozens of larger regions that are likewise grouped into three even larger, general *parts* of the brain. The lowest (both physically and functionally) of those parts is the hindbrain, which sits near the top of the brain stem connecting to the spinal cord, right at the back and lowest part of our skulls. This part of our brain is evolutionarily the oldest and controls those basic and automatic bodily functions we spoke of earlier, such as breathing, heart rate, blood pressure, digestion, and body temperature.

The second major part of the brain, and the second to evolutionarily grow in complexity, is the cerebellum. Located above the hindbrain, it plays a primary role in balance, motor control, and certain cognitive functions like attention, basic categorizing skills, and primordial emotions like fear and pleasure.

The third major part of the human brain is the cerebrum. It comprises 75 percent of the brain by volume and 85 percent by weight. The outer layer of the cerebrum is the cerebral cortex, where things happen that are truly unique and define us as human. The cerebral cortex is essentially a sheet of brain tissue about the size of a large deep-dish pizza. This part of the brain only exists in mammals.[9] And although some other mammals have brains larger than ours, the human cortex is more tightly folded into ridges wrapping around the rest of our brain, thereby giving us a much greater surface area that's still able to fit inside our heads. This is where memory, reasoning, problem-solving, perceptual awareness, speech, and our own human-level thought and, we believe, consciousness occur. It is, essentially, the seat of our intelligence and, as you might recall from Chapter 2, the advanced development of our neocortex (meaning "new" cortex) was the evolutionary enhancement that provided us, *Homo sapiens*, with our greater intellectual abilities over the other human species before us.

There's a misconception, which has now become common myth, that we use only 10 to 20 percent of our brain. In truth, we use virtually all of it—and most of our brain is active most of the time. Brain scans show that no matter what we are doing or thinking, all areas remain relatively active and none are ever completely dormant or shut down. Even when we're sleeping, all parts of our brain show at least some levels of readiness and interactive activity.[10]

For all of its complexity and despite the continual engagement of all of its parts, the human brain is extremely efficient in energy consumption. Requiring only 20 watts, which is barely enough to light a dim incandescent lightbulb, it's about 50 million times more efficient than any of today's computers of even remotely comparable capacity.[11] That's fortunate for us, because we can only produce a certain amount of energy from the volume of food we're capable of eating on any given day. Still, despite its efficiency compared to its total energy consumption, our brain does demand more energy than any of our other organs. Although it only weighs about 2 percent of our total body mass, it requires almost 20 percent of the total energy we generate. At that rate, the brain consumes 10 times its pro rata share of our available energy, or approximately 500 of the 2,400 calories we consume on an average day.[12]

There's good reason for its disproportionate energy consumption. Unlike computers that generally dedicate all of their resources to whatever single computing task they're executing, our brains must constantly manage a multitude of tasks, and a considerable portion of the effort must always be dedicated just to keeping us alive. That requires continual unconscious monitoring and regulation of our internal organs and a never-ending assessment of the environment around us. The critical processes that keep our bodies operating are remarkably complex. It takes a lot of caloric energy to manage our automatic bodily functions, including the constant monitoring of our five senses.

Beyond those basic managerial functions, the process of creating a single thought, regardless of how simple or complex it may be, also takes tremendous energy and vast biological computing power. All of our intellectual thoughts foundationally rely upon memory. Everything we consciously think and emotionally feel is based upon the experiences and information we've accumulated throughout all of our prior experiences and information. To accomplish the task of analyzing the onslaught of information constantly flowing into the human brain, and to do it through a continual search-and-compare process an uncountable number of times per second, our network of neurons must all work simultaneously to discern relative meaning and significance from anything that might trigger them into responsive firing.

Without memory, it's definitionally impossible for any individual, animal, organism, or other structure—computers included—to learn. It's just not possible. And that's true whether the information stored and recalled from memory was first acquired from actual experience or some other way. Many mysteries remain about how neuronal brain cells work together to produce, store, and access memories. Unlike computer memory, which we'll discuss below, individual human memories don't reside only in localized, specific spots. Instead, the recollection of an individual memory, which is a stored personal experience or a collection of otherwise acquired information, results from a dynamic, interactive process of numerous neurons firing, as a network, back and forth, in reaction to a stimulus.

Neuroscientists have identified five primary types of human memory and have roughly outlined the general locations where they occur. Short-term memory, or *working memory,* is the information that's only maintained momentarily to accomplish the task we are currently engaged in. It's the type of memory that's needed for instantaneous activity with respect to the requirements of the moment. As an example, the spatial characteristics of stairs we might be climbing or the ground over which we might be walking are types of information that our brains turn into short-lived working memory in order to coordinate and manage each of our physical steps. Once we've taken those steps, however, the information can be discarded since it will never again matter to us.

Our long-term memory is stored and activated essentially in the same manner as working memory, through a collective network of contributing neurons. The difference is that long-term memories are more localized to specific areas of the brain. *Episodic long-term memory*, the ability to remember the details of specific events, can last for a lifetime and seems to be located primarily in the hippocampus, below the cerebral cortex. *Semantic long-term memory*, the ability to learn facts and relationships, is stored in neurons located mostly within the cerebral cortex. And, finally, *instrumental long-term memory*, which gives us the ability to modify our behavior by recalling rewards and punishments, resides mostly in our cerebellum.

Because of the way the human brain stores information and memories between neurons in different areas, it was difficult until recently to determine or even reasonably estimate how much total information the brain can hold. Now, though, scientists have determined a means of doing so, and they've concluded that our brain's capacity is far greater than ever before thought. It's now believed that each one of our synapses can hold, on average, the rough equivalent of 4.7 bits of information.[13] That means that with its 86 billion neurons, each of which contains thousands of synapses that can effectively re-create memories through relations with other neurons, the human brain has the capacity of something on the order of 2.5 petabytes (2.5×10^{15}), or 2,500,000,000,000,000 bits of total storage capacity. That's the equivalent of 20 million four-drawer filing cabinets filled with solid text documents, 4.7 billion books, or 670 million web pages![14]

So, despite what was once thought, the amount that any human can theoretically learn is not limited in any practical way by the brain's storage capacity. We're all physiologically capable of remembering far more than we could ever possibly need. There are, however, a number of other biologic and practical factors that do limit how much we can learn or ever individually know.

The first of those factors is our limited scope of attention. Despite all of the information that flows into our brains every instant of our existence, even when we're sleeping, we can only pay conscious attention to a small amount at once—we have limited bandwidth. And since focused,

conscious attention is generally necessary to create long-term memories, much of the information that passes into our brains is never actually logged into long-term memory but, instead, simply disappears—in
one proverbial ear and out the other.

The second factor that limits our practical ability to learn is the order,
or timeline, in which we acquire our information and, hence, the strength
with which it resides in our memory. As humans, the things we learn
first tend to be embedded strongest in our memories, whereas something we learn later—particularly if it's contrary to an earlier impression
or understanding—tends to be weaker. It turns out to be true that it's
comparatively more difficult to teach an old dog new tricks . . . and
that's due to the strength of established patterns, practices, and beliefs.
Once we've learned to think a certain way or believe a certain thing, it
becomes more and more difficult to think or believe differently.

Another problematic factor of human memory is that we forget
things rather easily. Our memories deteriorate over time, particularly if
they weren't significant or considered by us to be important when they
were first logged, and also if they've not recently been recalled. As time
passes by, our ability to fully and accurately recall older information and
memories just naturally becomes increasingly more difficult.

Perhaps the biggest limitation of human memory, however, is the
one that's most obvious—we only have access to our *own* memory. The
amount of data or information we've somehow acquired is all we have to
call upon. Excluding our cell phones and other digital devices, there's no
way, at least currently, for us to biologically connect or combine the information in our own memories with the memory banks of other people.

———

Just as with humans, at the core of all computers is their memory—and
without memory, there can be no artificial intelligence or machine learning. Somewhat similar to a brain, total memory in a computer consists of
two parts. The first is RAM, which stands for Random Access Memory.

This is where information, or data, is stored that's roughly equivalent to the brain's short-term, working memory. RAM is used by the computer's central processing unit (CPU) for information that it needs to access and use very quickly, like the fundamental steps or instructions in a program. This information isn't stored permanently. Instead, RAM memory resides on the RAM circuit board or chip only momentarily in memory cells that lose their electrical charges very quickly. They therefore need to be constantly refreshed, or electrically boosted, by the electrical power flowing into the computer in order to maintain the information long enough to execute the task for which they're required. If power flow into the memory cells is lost, then the short-term RAM memory is lost as well. That's why almost all of a computer's power consumption is directed to the RAM operation. That's also why, when your computer shuts off, you lose all of the information that you or your program didn't previously store to long-term memory.

Today's computers run quite smoothly on 8 GB (gigabytes) of RAM, but 64-bit operating systems (explained in the last chapter) can easily handle 32 GB or even 64 GB of RAM, with the advantage being that the more RAM space that's available, the quicker and more efficiently the system can reach into and out of long-term memory banks as it steps its way through instantaneous programming requirements to produce seamless results. Because these speeds are all so fast in comparison to our human scales of perception, the added benefits of increased RAM are usually only noticeable to us in video displays or video streaming rates—where we can actually *see* the results of how smoothly a high-speed programming action occurs. This is why increased RAM capacity is especially important in today's highly developed gaming applications.

The long-term memory bank on a computer is known as the hard drive. This is where all permanent information is saved on the computer along with all software programs that are installed. When you type the words of a document on the keyboard of your computer, all of the onscreen changes are accomplished through RAM—which has pulled your word processing program from long-term memory, along with its

ASCII codes (explained in the last chapter), into the RAM memory cells so the instructions can be performed quickly. However, once you hit the save button, whatever you've created is then transferred and forever saved to your hard drive's long-term storage.

The long-term memory advantages that computers have over humans are twofold. First, once information has been stored, it deteriorates far slower than biologic memory. As long as the computer has been maintained, hasn't crashed, and has access to the location in which the data is recorded, it can generally be retrieved. Even information that may not have seemed relevant for a later analysis, comparison, or purpose when it was originally stored can nonetheless be subsequently accessed and evaluated in full and precise detail.

The second distinct advantage computers have over us is that they can connect and immediately share all stored information and memory between them. Whether it's through localized networks, the internet, or the cloud (all of which we'll discuss in Chapter 8), computers can instantaneously access the entirety of all data that's retrievable from any storage system or network to which they have access. That access— coupled with the ability of machine learning programs to meaningfully analyze all available data through artificially intelligent applications— is the single most profound implication of computing powers and AI capabilities going forward. Just imagine the intellect and insight any one of us would have if we could immediately analyze all of the information and experiences stored not only in our own mind, but in the minds of all our human colleagues. Machines that can access all of the world's retrievable information, instantaneously analyze and learn from it, and then provide us with new answers, strategies, and solutions are changing the realities of our existence and allowing us to accomplish far more than humans alone ever could.

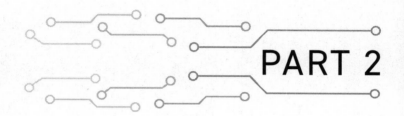

PART 2

TWENTY-FIRST-CENTURY COMPUTING AND AI

Power, Patterns, and Predictions

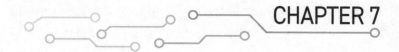

CHAPTER 7

GAMES MATTER

A man will be imprisoned in a room with a door that's unlocked and opens inwards; as long as it does not occur to him to pull rather than push.

—Ludwig Wittgenstein, 1889–1951
Austrian British Philosopher of Language, Logic, and Mathematics

One of the fundamental realities of human life is that we must compete to get along. It's a necessity that dates back to the earliest days of our evolution and to our most primitive struggles for survival. Like all animals, we must continually vie for resources, partners, and opportunities. Even in today's modern world, many aspects of our lives are shaped by unavoidable competitive factors. Rivals are all around us. Intellectual, social, economic, and even physical contenders, or at least the potential threat of them, are everywhere.

But the reality of our competitive essence goes even deeper. It emanates from the very physiology of who we are as a species. At a biological level, we are always competing against the physical circumstances and situations that every new moment presents. Our brains are constantly gathering information, evaluating it for significance, adjusting our attention, and coordinating our reactions however necessary. Every instant of our existence is a new learning experience and another necessary performance. Just leaving our homes to go to work or to a park requires a steady assessment of new and constantly changing information. To complete whatever steps we take throughout the day, whatever tasks we undertake, strategies we select, and goals we accomplish—all of it requires incredible analysis and competitive efforts along the way. It's us against the world.

As much as competition is a fundamental and unavoidable necessity of human life, it's also something we eagerly seek out. We all love to play competitive games. From the time we're toddlers, games entertain, occupy, and amuse us. They also teach us many of the fundamental skills by which society will measure us. When we're young, games ripen our minds, train us to analyze and react decisively to changing situations, and sharpen our ability to focus in order to achieve our goals. They provide a ready means of measuring ourselves against others by assessing our weaknesses and developing our strengths. We learn to operate within the boundaries of established rules, but we're also able to practice stretching those rules, strategically, to our best advantage. As we grow older, games continue to entertain and challenge us, while also keeping us mentally, socially, and even physically fit. Most importantly, though, games provide all of those benefits in contexts that are not only enjoyable, but also relatively cost free. For those of us who don't earn our income or livelihoods playing games, the risks of recreational competition are comparatively small . . . and are far outweighed by the benefits we gain.

———

Games fall generally into one of four categories. There are games of chance—like roulette, lotto, and dice—where luck is the only determiner of outcome. There are games of pure intellect—like crossword puzzles, brainteasers, checkers, and chess—in which we match our knowledge, skills, and strategies against a set standard or an opponent. There are games of pure physical aptitude—like individual track-and-field events—where success is determined only by the participant's own physical attributes. And, finally, composing the largest category of all are games that combine two or more of those elements. Most team and sporting events fall into the last category, where we test our skills and strategies against a combination of elements outside our control—like the skills of an individual opponent, the aptitude of an adversarial team, the luck of the draw, and even the bounce of the ball.

Games come and go, but those that last longest—over decades and even centuries of time—reflect much about the priorities and values of the cultures that continue to play them. Invariably, the games that survive from generation to generation teach and require skills that the cultures behind them deem relevant and worthy of reward. And as games survive through the ages, the styles and strategies that are handed down reflect the best skills and cumulative analyses of the players that came before.

Some of the most popular and long-lasting games of all are played on boards—with move-based rules that ensure each player stands an equal chance of winning. Certain board games showcase the most unique strengths of the human mind, including our abilities to analyze, anticipate, and strategize. Chess is a perfect example. It's a game that's been popular for centuries, across many cultures. And, for a variety of reasons we'll soon see, it eventually became the ideal platform for the early research and development of AI.

———

Although the exact origin of chess isn't definitively known, its history can be traced back more than 1,500 years. The game originated in India, in either the fifth or sixth century CE, and it's believed the original rules incorporated significant elements of chance, with dice rolls determining the permissible movement of the pieces.[1] But as the game spread to Persia and then throughout all of the Arab world, it gradually evolved into a strategies-only game, with moves based solely on rules that remained constant and equal for each player throughout the game. As the rules continued to evolve, it spread to Europe sometime in the ninth century, where it first became popular with the nobility and upper class, and then with the lower merchant classes as well.

Throughout all the centuries that chess has been played, the greatest players, the grandmasters, have always been held in high esteem and commonly regarded as true geniuses. Although chess is a game mastered only by the most intellectually acute, as far back as the 1500s people began imagining a day when machines might also be able to play chess — on their own, autonomously somehow, without human direction.

Such ideas were only imaginative and provocative *what-if* thought experiments until, in the early 1770s, a Hungarian inventor named Wolfgang von Kempelen constructed a mechanical "automaton," a machine he called *the Turk*. Von Kempelen gave his intricate machine an artificial face, body, arms, and hands. He dressed it in robes and a turban, and proceeded to convince people throughout all of Europe that his creation was capable not only of beating the very best humans, but even of moving the pieces around the board itself.[2]

For decades, von Kempelen's hoax successfully deceived his audiences. He convinced commoners, academics, early scientists, and aristocrats alike that his eerie, humanlike machine was capable of conscious deliberation, decisions, and movement. The reason the deception succeeded was that observers couldn't see through the disguise and obstructions von Kempelen had constructed. Masterfully, he had designed the Turk with enough hidden space inside its main cabinet

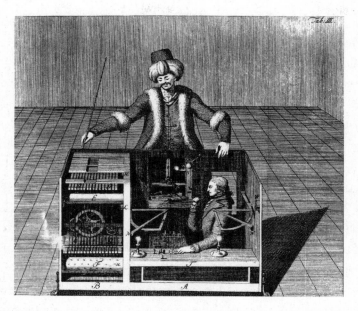

Figure 7.1: Copper engraving from Karl Gottlieb von Windisch's 1783 book, *Letters About the Chess Player of Mr. von Kempelen, Along with Three Engravings Depicting This Famous Machine*, in which von Windisch speculated about the mysteries of the Turk.

to conceal a small person who secretly operated the movements of the machine. See Figure 7.1. Unaware that a skilled chess player was hidden within, observers naively believed there must be some magic or mystical influence at work. They intuitively sensed, even then, that a device made only of metal gears and wires couldn't just radiate consciousness without some otherworldly intervention. Although many people proposed theories of how the Turk operated, von Kempelen never admitted or denied any of them, and his secrets remained a well-guarded mystery for over a hundred years.[3] It was one of the most successful illusions of all time.

Two hundred years after von Kempelen's creation, when digital computers were first being seriously developed in the 1960s, a machine capable of playing and beating a human at the highest level of chess was still the holy grail of computer capability. The history of how that goal was eventually accomplished, and then surpassed, is an enlightening tale of human competition and innovation. It also illustrates the earliest steps toward artificial intelligence and how machine learning eventually evolved.

———

Modern international chess tournaments began in the second half of the nineteenth century, and the first unified World Chess Championship was held in New York in 1886.[4] From the start, the title of World Chess Champion was considered a symbol of cultural, national, and even political esteem. The first world champion was Wilhelm Steinitz, a Bohemian Austrian who had emigrated to New York a few years before the world match. While not yet an American at the time of that first world tournament, Steinitz nonetheless insisted on placing the American flag next to him throughout the competition, a move that immediately framed the championship as an important source of national and ideological pride.

Two years after winning the first championship, Steinitz did become an American citizen, and he went on to hold the world title under the American flag for a total of eight years, until Emanuel Lasker captured it for the German Empire in 1894. Lasker then held the title under Imperial Germany's flag for almost three decades, until his 1921 loss to a communist Cuban, José Raúl Capablanca. In turn, Capablanca lost the title six years later to a Russian named Alexander Alekhine.[5]

Although Alekhine was a Russian subject by birth, he opposed the new Soviet regime that had taken power after the Russian revolution. After winning his championship, he renounced his Russian citizenship as a show of political protest and emigrated to France. Alekhine's denunciation of Soviet Russia infuriated the regime, especially as Alekhine went on to hold the title (with only one short interruption) for 25 years, all in the name of France. Eventually, after the end of World War II, Alekhine agreed to defend his world title against a heralded Soviet player named Mikhail Botvinnik—but a curious situation would soon unfold.

In 1946, while Alekhine was at a secluded location in Portugal preparing for the upcoming match, he was found dead in his hotel room. The autopsy report indicated the cause of death was accidental choking, but the circumstances were extremely suspicious and few believed the report's findings. It was widely believed that either the French Resistance was responsible for killing Alekhine because of his alleged Nazi sympathies or, alternatively, that the Soviets killed him simply to avoid the culturally embarrassing possibility that he might beat the Soviet challenger, Botvinnik. In any event, no charges were ever brought, and the cause of Alekhine's death is still questioned even today.[6]

Following Alekhine's untimely demise, a tournament was held to fill the vacated title. As most expected, Botvinnik won the championship in easy fashion. With the trophy finally in hand, the Russian authorities unabashedly promoted their title as proof of the cultural and intellectual superiority of communist Soviet society over the decadent, less disciplined ways of the West.

For 24 years, the world title was held by Botvinnik and his three Soviet successors until it was lost to an eccentric young American named Bobby Fischer, who defeated the USSR's Boris Spassky in 1972. The match between Fischer and Spassky was played over the course of two months in Reykjavík, Iceland, and was publicized as the match of the century . . . a Cold War confrontation between the US and USSR. More than any game or sporting event before it, the Fischer-Spassky match was hyped by the media as a metaphor for the political times, fraught with international intrigue and interest. It lived up to its billing, gained worldwide attention, and played out with one controversy after another—especially because of Fischer's notoriously difficult and disagreeable ways. Eventually, though, Fischer prevailed, and thereby became the first American-born player ever to hold the championship.[7]

But Fischer's reign only lasted until 1975, when he refused to defend his title because of a dispute with chess's international governing body. Fischer's refusal to play resulted in a forfeit, and the title was returned by default to the Soviet Union, this time to a player named Anatoly Karpov.[8]

Karpov quickly became a Soviet national hero and proved himself a worthy champion by repeatedly defending his title for the next ten years. Rising in the wings, however, was another young Russian who would ultimately become the most recognized chess player of all time, a culturally iconic figure destined to be forever linked to the power of computers and to the public's first impression of artificial intelligence. His name was Garry Kasparov.

It took Kasparov, the world's strongest contender for Karpov's title, two long and drawn-out matches to eventually wrestle away the championship. The first Kasparov-Karpov match was held in 1984 and lasted a grueling 48 games that took more than five months to play. The match was eventually terminated, without a winner, over concerns for the players' exhaustion and health.[9] A rematch was scheduled for the following year, this time with a revised format to ensure an actual winner in no more than 24 games. The second match was another long struggle. It lasted two months and came down to the final game, from which Kasparov finally emerged the victor and new world champion.[10]

At only 22 years old, Kasparov was articulate, charismatic, and likeable. People everywhere enthusiastically embraced him as the new face of international chess. He quickly became a popular celebrity throughout the world, and he held on to the championship until 2000.[11] In the years ahead, though, all of Kasparov's fame would skyrocket not just because he was an international figure, but because he would become known as the man to make humanity's last stand against technology.

———

Kasparov's epic battle against computer technology took place in May 1997, while he was still the reigning world champion. He was in his mid-thirties, and the eyes of the world were upon him. A match was scheduled that pitted Kasparov against the most powerful computer ever built, the IBM supercomputer RS/6000 SP. It was contained within two enormous black metal towers, weighed 1.4 tons, and was dubbed Deep Blue.[12]

The match would be the second between Kasparov and the machine. In the first, 15 months earlier, Deep Blue was much slower and less sophisticated, and Kasparov had come out the victor. This time, though, the world sensed that things might be different. Although Kasparov won the first match, things hadn't gone quite as easily as expected. He had won by a score of 4–2 overall, but the monstrous machine had exposed some of his human vulnerabilities. Also, IBM's team of engineers responsible for designing and programming Deep Blue had learned from their shortcomings in the first match. Over the following year, they doubled Deep Blue's computing power and speed, added chess grandmasters to their team of computer scientists and programming experts, and invested untold hours, expertise, and dollars to further develop their computer. IBM had pulled out all the stops, with only one purpose—to defeat Kasparov, regardless of the cost. Prior to the rematch, the leader of the Deep Blue team said, "I anticipate that we will win this match, overwhelmingly."[13]

Deep Blue didn't utilize artificial intelligence in the way we think of it today. Instead, the machine relied only upon its incredible computing and calculating power. Chess is an extremely complicated game, not because of the complexities of the rules or the different pieces and allowable moves, but because of the vast number of alternative countermoves that are possible at every step. At each move, a player's strategic decisions must anticipate the opponent's likely responses and foresee as many paths as the game might possibly take. Players with average recreational chess skills can generally think only a few moves ahead, while grandmasters of Kasparov's talent can think up to 12 or more potential moves ahead. But even that level of analysis and strategic planning is incomplete, shallow even, compared to all of the possible variations that can occur over the course of just a few moves in any chess game.

At any one point in the alternating turns between chess players, there are an average of 30 possible moves a player can make, and from each of those moves there are 30 possible countermoves his or her opponent can likewise make, and so on. Of course, not all of those moves would be strategically sensible, and they're therefore not all worth serious evaluation. Many, however, are. To put things in context, after only the first move by both players in a chess game, there are 400 different board setups that could possibly occur. After both players have moved a second time, that number escalates to over 197,000. And, after each player's third move, more than 120 million game variations could already have occurred. Finally, after 40 moves, which is the length of an average chess game, there are something on the order of 10^{120} possible unique game variations that could have been played. From our discussion back in Chapter 3, you can imagine the absurd enormity of a number like 10^{120}. By comparison, it's estimated there are *only* 10^{15} hairs on all the human heads in the world, 10^{23} grains of sand on all the beaches of our planet, and 10^{81} atoms in all of the observable universe.[14] The total number of possible combined board variations, in any one game, is more than all of those numbers combined. As a practical matter, that means that not long into *any* chess game, the competitors are often playing a game that's different from any game ever played before.

Given that computers have a distinct advantage over humans in analyzing such a tremendous number of potential variations, chess software programs are generally written to evaluate moves—mathematically and methodically—in an expansive search tree. In theory, they examine all possible moves, then all countermoves to those moves, then all moves countering them again, and so on. For each step, the program essentially assigns a strength evaluation to each of the available next board positions. This calculation continues until a certain maximum search depth is accomplished, depending on the computer's power and the program's depth of sophistication, and then the strongest next move is selected.

By 1997 standards, Deep Blue was immensely powerful—capable of making 200 million calculations per second,[15] all of which were designed to evaluate the various positional strengths on the board. It was capable of analyzing up to 30 moves ahead, with each of those possible moves assessed by mathematically evaluating the various search trees they generated and then determining, statistically, the best move to make. Deep Blue, simply put, calculated probabilities in an extraordinarily powerful way. It played by brute force at a level that was far more advanced and impressive than any machine before it.

At the second Kasparov–Deep Blue match, played on the 35th floor of IBM's New York City headquarters over the course of nine days in May 1997, the likeable Russian was on a stage unlike any other of his career. From his perspective, the ultimate purpose of playing Deep Blue was altruistic. Although a $700,000 winner's share of the $1.1 million purse was at stake, Kasparov saw his efforts as a genuine, cooperative contribution toward developing the state of technology. He came from a scholarly background and his mother, who was a teacher, had greatly influenced his attitudes. In his mind, the match against IBM was more than just an income opportunity. It was an academic exercise that would ultimately prove valuable for everyone. With that said, Kasparov also had unwavering confidence in his abilities. Before the match, he claimed, "I don't think it's an appropriate thing to discuss the situation if I lose. I never lose. I have never lost in my life."[16]

But most of the world was more concerned than Kasparov and felt that a more fundamental, humanistic risk was at stake in the match. Television, newspaper, and magazine media from around the globe flocked to New York, and although their readers, listeners, and viewers were uncertain what would happen, they were overwhelmingly eager for another Kasparov victory. They didn't consider his role a cooperative effort to advance the state of computer technology or brute-force computing capabilities, and they certainly didn't recognize that, in many ways, the match was a carefully orchestrated undertaking by IBM to increase the quality and value of their intellectual property. Instead, most everyone saw the match as Kasparov's calling to defend the integrity of human intellect and to validate the superiority of the human mind over a mechanical fabrication. It was Kasparov's responsibility to fend off the onslaught of technology. He wasn't playing for himself or for Russia. He was playing for the world, for the species. And even if it was inevitable that computers would eventually surpass humans at our most iconic strategy-based game, most of the world believed, or at least deeply hoped, that the task IBM faced was still far from achievable.

Headlines around the world cheered the human flag carrier on. The week before the match, the cover of *Newsweek* featured a stern, penetrating photo of Kasparov and in large print declared the upcoming event, "The Brain's Last Stand."[17] The related article warned, "We stand at the brief corona of an eclipse—the eclipse of certain human mastery by machines that humans have created. How well Kasparov does in outwitting IBM's monster might be an early indication of how well our species might maintain its identity, let alone its superiority, in the years and centuries to come."[18]

The first game of the match between Kasparov and Deep Blue started off slowly, with Kasparov playing conservatively—probably to get a strategic feel for how well the new version of Deep Blue could play. Early on, the computer adopted a much more challenging and aggressive style than Kasparov. But, toward the end of the game, it made some perplexing moves that weren't of any competitive value. Those moves, coupled with the usual, strong play of Kasparov, gave him a surprisingly easy victory. At the press conference afterward, a packed auditorium greeted Kasparov with relieved and uproarious applause. His confidence was in full display, and people around the world let go of their fears. It seemed apparent, at that stage, that biology would again triumph over technology. Standing at the same press conference, however, the IBM team was completely unshaken . . . as though they knew something the rest of the world didn't.

In Game 2, things took an entirely different turn. This time, Deep Blue departed from the aggressive style of play it had used in the opening of Game 1, and instead employed a slower, more positional and developing style to match the same from Kasparov. Then, at a critical midway point in the game, the computer changed the dynamics with an extraordinary, conspicuous, and seemingly all-too-humanlike mistake. Instead of continuing the positional back-and-forth sparring that had been taking place, Deep Blue varied its strategy—for just one move, on its 37th turn—to block one of Kasparov's pawns from moving forward. It seemed to be a wasted and entirely unnecessary move at that stage, and looked to everyone watching, including Kasparov, like an error—the kind a human might make in a moment of distraction.

Deep Blue's 37th move wasn't a mistake . . . but instead a brilliantly foresighted strategy. Only a few moves later, Kasparov's inability to advance the blocked pawn seemed to paralyze his chances to win. For Deep Blue to have blocked the pawn, at the point in the game it did, was the kind of strategy, the kind of creativity, that no one had seen from a computer before. Sure, the move was one of the options the machine evaluated by the brute force of 200 million calculations per second, but the move wouldn't—or shouldn't—have scored high on the list of best options. It seemed a very strange choice for the computer to have made. And when it later proved crucially effective, Kasparov was so astonished and dismayed that he eventually resigned the game, thinking it was impossible to avoid defeat. In truth, and to make matters worse, analysis after the game showed that Kasparov still could have forced the game to a draw, even despite Deep Blue's infamous 37th move. Kasparov's psyche was crushed. Like many watching, he was convinced that IBM was cheating and that one or more of the human grandmasters IBM had recruited to their team must have been working with the computer, behind the scenes, during the course of the game. He threatened to quit. IBM denied they'd done anything wrong and the match went grudgingly on.

Distracted by the events of the second game, Kasparov's focus and confidence were gone. Games three, four, and five all ended in draws. The human was doing his best. No one could deny that. But Kasparov had become exhausted by his efforts, both mentally and emotionally.

With both machine and man having only won a single game, victory in the match hinged on the final, sixth game. But Kasparov played poorly in the end and resigned after only 19 moves. It was the shortest game of the competition and concluded the match. The machine had won by a score of two wins, one loss, and three draws. Kasparov was distraught and angry. He asked for a rematch, which in the world of championship chess would normally be granted, as both a courtesy and a routine. IBM, however, refused. They dismantled Deep Blue and dispersed its component parts to different locations. *That* machine would never play chess again.

IBM had proven the strength of their new technology and immediately reaped the benefits of the commercial exposure and worldwide publicity the event provided them. The day following the match, IBM's stock rose to a ten-year high and many estimate its corporate revenues were ultimately increased by billions of dollars as a result of the win.[19]

Although Kasparov was distraught by the loss and remained publicly agitated and angry for years, he has since gone on to become a major proponent of computer technology and a vocal advocate of its valuable role in advancing human knowledge and strategic thinking. He's also an ardent and enthusiastic supporter of artificial intelligence.

———

While Deep Blue surpassed the best that humans can offer in the game of chess, it did so only through the brute force of massive computing power and the tremendous number of calculations it could make. The human brain, however, does much more than construct extravagant search trees and calculate probabilities. As we discussed in the previous chapter, people think in the context of all of our experiences. We compare, contrast, interpret, and analyze everything we already know. In the process, we learn. We discover patterns to make sense of things, to make predictions, and to arrive at logical conclusions. To think and to learn more like a human, computers therefore needed to do more than Deep Blue. They needed to utilize their immense power of computation, but they also had to somehow move beyond the rote process of simple calculation. They had to evaluate information in broader, more interpretive ways. In pursuit of those steps, the next great breakthrough in accomplishing an artificial intelligence more like our own was also developed in the context of a game.

The game of Go originated in China thousands of years ago. Despite its relative obscurity outside East Asian culture, it's the oldest continuously played board game of all time.[20] In ancient Chinese, Japanese, and Korean cultures, Go was considered one of the four pillars of society and one of the four principal constructs of education—along with music, poetry, and painting. In Chinese, it's called Wei Qi, 围棋, which roughly means "surrounding game." In Korean, it's known as Baduk, 바둑. In Japanese, it's called Igo, 碁.

The game is very simple to play, but also incredibly nuanced, abstract, and extremely difficult to master. There is only one type of piece and only one type of move. It's played on a board that is 19 by 19 squares in size—by comparison, chess is played on a board that's 8 by 8. The goal of Go is to place your stones on any one of the 361 lined intersections of the board in an effort to create a linked group of stones surrounding the most combined open territory. When a player surrounds an opponent's stones, they're captured and removed from the board, rendering the spaces they've left behind part of the captor's overall territory. At the end of the game, whoever has the most territory wins.

Regardless of its simple rules, Go is incredibly complex. At most points in the game, the best players can't even begin to calculate in their minds all the moves that might *mathematically* be the next best. There are just too many options and possible game variations. A *sense* for the game is required, an *intuition*, a feel for the right move under the right circumstances at the right time. We noted that there are 10^{120} possible variations of *chess* games that can occur over an average 40-move game. That number is more than all the atoms in the universe. By contrast, in Go there are something along the lines of 2×10^{170} game variations that can occur,[21] a number that's unimaginably greater still. See Figure 7.2.

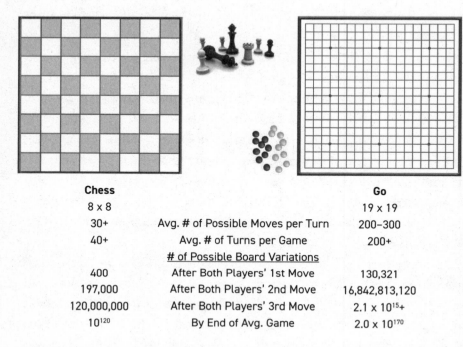

Chess		Go
8 x 8		19 x 19
30+	Avg. # of Possible Moves per Turn	200–300
40+	Avg. # of Turns per Game	200+
	# of Possible Board Variations	
400	After Both Players' 1st Move	130,321
197,000	After Both Players' 2nd Move	16,842,813,120
120,000,000	After Both Players' 3rd Move	$2.1 \times 10^{15}+$
10^{120}	By End of Avg. Game	2.0×10^{170}

Figure 7.2: A mathematical comparison of the numbers that characterize the games of chess and Go, and the number of possible game variations that can occur.

So, how can a computer deal with the exponentially greater mathematical complexities of Go? The answer is not by brute force alone. Computational power is insufficient without something more. Even the fastest supercomputer could never work through all the possible Go variations in any meaningful time frame.[22] The only alternative to brute force, then, is for a computer to evaluate the strategies of Go by thinking more like a human—in other words, to evaluate current conditions in the context of all of its *experiences*. And that's where modern machine learning applications first came into the picture.

The most significant step forward came from a London-based company called DeepMind, which was founded in 2010 by Demis Hassabis—a former child chess prodigy, digital game developer, and neuroscientist. The company began its research into AI by creating a learning algorithm that taught *itself* how to play common Atari video games from the 1970s and '80s. DeepMind's program was developed through games like *Breakout, Pong,* and *Space Invaders* without any prior programming of the games' rules or strategies. The program was only coded to understand that the goal of the games was to collect points in order to accumulate the best possible score, nothing more. By playing the games over and over again at computer speeds, DeepMind's program was able to learn very quickly from its own errors and mastered the games in a matter of hours—not months or years like a human would require. With the DeepMind team's research and algorithms continually proving themselves after learning to play early Atari games, the company was purchased by Google in 2014 for over $500 million.[23]

As the Google/DeepMind scientists and programming experts became more adept at what their machine learning programs could accomplish, they decided to tackle the challenges of Go. They began feeding data from thousands and thousands of human-played Go games into a new algorithm they called AlphaGo and allowed the computer to digest, sort, and analyze all of the information. The algorithm was then set loose to play against itself, again thousands upon thousands of times. Summarily, AlphaGo's approach to each new game it played involved three steps. First, it drew upon its ever-expanding bank of prior games that it had analyzed and played. Then it scanned the board positions of the game it was presently playing to find strategically promising board spots to consider as options for its next move. Then, through search trees significantly more complex than those Deep Blue had employed in chess—and more similar to the kind of analysis that human brains make—AlphaGo comparatively evaluated how promising the probable outcomes of each of those plays would be.

In essence, DeepMind was the first successful and widely recognized application of reinforcement machine learning—a process through which a computerized construction of hardware and software, called deep neural networks, mimics the web of neurons in the human brain, thereby allowing for an evaluative, self-improving process. In Chapter 9, we'll discuss neural networks and various other elements of machine learning in detail.

In 2015, the DeepMind team decided it was time for AlphaGo to challenge a human expert. They found a worthy and willing opponent in the European Go champion, Fan Hui, who was an extremely strong player, but still far below the most elite masters of the game. The match wasn't a contest. AlphaGo beat Fan Hui in five straight games with relative ease. Many in the Go community, however, dismissed the computer's victory, believing it was one thing for AlphaGo to succeed against Fan Hui, but that it would be quite a different, impossible task to prevail against a true master who played Go at the highest level of human intuition and imagination.

In response to the doubts, DeepMind brought Fan Hui onto its development team and scheduled a match for only a few months later against a South Korean player named Lee Sedol. Like Garry Kasparov in his chess match against Deep Blue almost two decades earlier, Sedol was the reigning world Go champion and had been universally acknowledged as the world's strongest player for well over ten years. He had 18 world titles to his credit.[24] He just didn't lose.

Once again, the match was seen as humanity's last stand against the unstoppable tide of technology and, once again, the world watched with rapt attention. In March 2016, media and press from across the globe scrambled to Seoul, South Korea, for a five-game contest that was scheduled over seven days. Reminiscent of the past, no one outside of DeepMind gave AlphaGo a chance. Consistent with a habit of underestimating AI developments that continues even today, most everyone, including the majority of AI researchers, believed machine learning technology was at least ten years away from succeeding at the challenge in store for DeepMind. Sedol himself said before the match, "I am confident about the match. I believe that human intuition is still too advanced for AI to have caught up."[25]

Played at Seoul's Four Seasons Hotel, the first game was broadcast live and watched by an estimated 80 million people worldwide, 60 million in China alone.[26] From the very start, it was clear that things weren't as people had expected or hoped. Beginning with the first move, AlphaGo played with an apparent creativity so different from conventional human playing styles that its moves seemed random at times, even outright wrong. But its new strategies quickly proved powerfully, almost intuitively, effective. In the words of one commentator, "No matter how complex the situation, AlphaGo plays as if it knows everything already."[27]

Sedol was surprised and immediately confused by the unexpected level and style of AlphaGo's play. He lost the first game in an overwhelming defeat. The world was shocked, especially the large community of Go players and fans throughout Asia. Worse, they were completely, culturally unprepared for the loss. In a single game between man and machine, centuries of methodically developed and highly respected theories of how Go *ought* to be played had been dismantled by the machine's drastically new strategies.

The second game went no better. Sedol lost soundly again. In the press conference afterward, he was clearly shaken by the incredible pressure upon him. The internet and media were alive with reactions, and most observers empathized with Sedol's plight. There seemed something very unsettling and sad, frightening even, about a machine that could find new and devastatingly effective strategies on its own . . . especially when those strategies were far different from anything the best human players, over centuries of time, had ever even thought to consider.

After another loss in Game 3, things had become entirely hopeless for Sedol. Having already mathematically lost the match, he publicly apologized for being powerless against AlphaGo. The press, sympathizing with his situation, reported that the best human player of our most complex and abstract game was simply without any meaningful weapons to wield against his computer opponent. Sedol stood completely exposed, fighting a cold and losing battle against a nemesis that didn't even exist in physical form.

In Game 4, with the match already lost overall, things were again looking grim for Sedol. To everyone's surprise, though, he made a creative move late in the game that was not only unconventional but beyond any serious player's reasonable imagination. Even AlphaGo's own game logs later showed that it evaluated the move as a 1 in 10,000 probability.[28] But with that single, unexpected placement of a stone, Sedol completely reversed the dynamics of the game and managed to eke out a win over the computer. The move was later dubbed "the God move." After the game, Sedol was greeted with euphoric applause by the assembled gallery of humans attending the press conference. Clearly relieved of a tremendous weight, Sedol said, "People felt helplessness and fear. It seemed like we humans are so weak and fragile. And this victory meant we could still hold our own. As time goes on, it will probably be very difficult to beat AI. But winning this one time, it felt like it was enough. One time was enough."[29]

The match ended with another win for AlphaGo, resulting in a final tally of 4–1 in favor of AI over humanity. At the end of it all, the style of AlphaGo's play upset centuries of accepted wisdom. The huge community of Go players and enthusiasts throughout Asia collectively reflected on what had occurred . . . and ultimately found inspiration. Their game had been played for thousands of years and was symbolic of their cultural way of life. It represented the wonders of the human mind and our unique, human way of mastering the challenges of the world around us. Yet, the game had just been creatively mastered by a machine. The community realized, though, that this was an opportunity for them to see things, perhaps all things, in a new light. After the match, Sedol said it most poignantly: "What surprised me the most was that AlphaGo showed us that moves humans may have thought are creative were actually conventional."[30] Fan Hui, the European champion who had been the first to lose to AlphaGo, also reflected, "Maybe AlphaGo can just show humans something we never discovered. Maybe it's beautiful."[31]

Following the match, DeepMind's team continued to improve their algorithms. In late 2017, they introduced a new version of the program, called AlphaGo Zero, that learned the game so well—by playing 4.9 million games against itself in only three days—it was able to beat the prior version 100 games to none.[32] The difference this time was that it learned *without reviewing any human-played games at all*. It was only given the rules of the game and was left, from the very start, to discover the best ways to win entirely on its own, thereby eliminating all human influence.

———

The next leap in developing effective AI from gaming platforms occurred in transitioning machine learning algorithms from playing board games to modern video games—and the explanation of how that was accomplished completes the significant bridge that led to many of our current AI applications.

Today's video games are incredibly complex, particularly those that are played online in multiplayer formats. Billion-dollar companies routinely grow from them and, since the early 2010s, professional teams from across the world have been competing in matches and tournaments organized around them. Top professional players earn multimillion-dollar contracts and the commercial values of teams now rival traditional professional sports franchises. Broadly known as *eSports* (electronic sports), competitions are sponsored by huge corporate advertisers and have become prominent, international multimedia events. The top tournaments award prize money totaling millions of dollars that's generally split amongst the winning team owner and players. Tournaments are streamed live over the internet and simulcast on television and cable networks—with reporters, analysts, and commentators just like traditional sporting events. Audiences for some of the biggest matches reach well into the millions, and eSports will even be a part of the 2022 Asian Games, and possibly a part of the 2024 Olympic Games.[33]

The two-dimensional playing surfaces of board games like chess and Go are replaced in modern video games by complex, three-dimensional virtual landscapes that players must navigate to accomplish whatever goals the games reward. When playing alone, a single player competes against the game itself, generally trying to accumulate points to reach the next level of the game. In two-player games, opponents compete against each other, either directly or indirectly, and victory is usually achieved by eliminating or scoring more points than their competitors. In multiplayer games, a number of people participate, trying to win either as individual players or as teams. In all of those formats, artificial intelligence algorithms are frequently used to generate automated nonplayer characters (NPCs), or creatures, within the game that battle independently against the human players by adaptively responding to the human-controlled characters as the game progresses.

The primary difference between video games and traditional board games is that many computer games, like life, aren't strictly turn based. Players have to make decisions, act, and react at the speed of immediate thought. Long-term strategies and considerations are required, but there's no time to calculate the next move, huddle, theorize, or deliberate. What a human player learns about his or her opponents' playing characteristics as the game progresses has to be applied, in real time, as part of the reactive equation. Moves and actions elicit immediate countermoves from the opposition. Also, players aren't always able to see what their opponents are doing when, for instance, they're on a different part of the game "map." As a result, things can happen unexpectedly, and moves can cause unforeseen, unanticipated peripheral consequences.

———

The most common and popular eSports games are multiplayer battle and fighting games. One of the most popular is *Dota 2*, a battle game produced by an American video game company called Valve Corporation. Played online and in real time between two teams of five players, the game requires each team to occupy and defend its own base within the virtual landscape. Each of the ten players independently controls his or her own game character, called a hero, and each hero has its own unique abilities, strengths, and playing characteristics. During a game, players attempt to collect different virtual resources and game points to help them defeat the opposing team's heroes in direct player-to-player virtual combat. The winning team is the first to destroy the other team's home base.

The dynamic playing condition of *Dota 2* was a perfect environment for the continuing evolution of AI and machine learning applications. OpenAI is a nonprofit company originally founded in part by SpaceX and Tesla's Elon Musk. Its expressed corporate mission is to ensure that artificial intelligence remains safe and benefits all humanity.[34] Throughout 2018 and 2019, the company engaged in a string of *Dota 2* challenges by matching a team of artificially intelligent learning algorithms, or *bots*, it had created against teams of human players with increasingly better skill levels. The OpenAI bots, as we'll discuss more in Chapter 11, don't have any physical form—they're only algorithms of code acting within the virtual space of digital software.

For the first developmental challenge, the OpenAI bots—named the OpenAI Five—were matched against teams of amateur human players. It didn't take long before those amateurs were no competition for the rapid learning abilities of the bots. So, OpenAI moved on. As its next step, the company challenged a team of midlevel, *professional* human players. In a best-of-three match, the OpenAI bots again won convincingly. A month later, OpenAI again increased the level of its competition by going against a top professional Chinese team. The tournament was a major international media event that drew millions of online spectators from around the world. Although the computer lost the two games it played against the Chinese professionals, both of the games were competitive—*Dota 2* games usually last 35 or more minutes, and the OpenAI bots maintained a good chance of winning for the first 20 minutes or so of each game.

Despite the loss, OpenAI proved that it had successfully developed an artificially intelligent system capable of quickly learning to manage the complexities and uncertainties of real-world competitive situations. Much like DeepMind's AlphaGo, the OpenAI bots had been trained by the trial-and-error approach of reinforcement learning. Instead of being coded with strategies devised by humans, or even with the rules of *Dota 2*, the bots were only given unstructured access to the game. They were thrown in, virtually, and left to figure things out for themselves. The bots started out knowing nothing about the game, not even the goal.

But in the dynamic, unpredictable, and constantly changing environment of the game, they learned on their own how to connect certain actions to rewards, and how to win. And, although the five bots acted independently of each other, they also learned to act in coordination, as a team, much like humans do.

Although the OpenAI bots lost to the top team of professional Chinese players, the only problem the bots seemed to have was their inability to recover once they fell far behind on points. In hindsight, the reason for that made sense. The bots had learned—in all of their prior games against only themselves and lesser competition—to pursue strategies just sufficient to win. By definition, only a small margin is necessary to achieve victory, and therefore only a slight advantage was all that ever *mattered* to the bots as they learned and developed their initial strategies. Because there was no advantage to winning by a large margin, the algorithms behind the bots had never discerned a strategic advantage to gaining a larger lead than necessary to ensure victory. And so, once they fell behind, they had difficultly devising the kind of coordinated response necessary to make up for large deficits. The bots had simply never been in that situation before.

In April 2019, the OpenAI Five again challenged a top professional team of *Dota 2* players, who this time were the reigning world champions. Although the match was scheduled for three games, the artificially intelligent team of virtual bots quickly proved that they had learned the value of crushing their opponent as quickly and decisively as possible. They so overwhelmed the human players in the first two games of the match that the event organizers didn't even bother staging the third game.

While OpenAI was developing its *Dota 2* algorithm, a Chinese company was likewise developing a way to ensure victory by an artificially intelligent team of bots, but it did so by playing a different online game, *StarCraft II: Wings of Liberty*. Like *Dota 2*, *StarCraft II* presents an ever-changing landscape of obstacles and various combatant characters that make the real-time challenges of the game incredibly complex.[35] One particularly realistic aspect is that human players aren't able to see all of their opponents' capabilities until well into the game, not until they've scouted the entire terrain, or map. This creates an inherent *fog of war*, which is a common reality of real-life competitive situations, including military encounters—where humans must always remain prepared for unknown strengths or strategies an opponent may be hiding or hasn't yet unveiled.

Beyond the inherent fog of war issue, *StarCraft II* makes matters even more difficult by including an AI feature that allows the game itself to cheat. It's actually called *Cheater AI*. At the highest settings, the game has the advantage of always knowing the locations of all characters, resources, and changing obstacles. Unlike humans, it doesn't have *any* fog of war. It's essentially omniscient and can account for and analyze all relevant elements both on and off the map. Understandably, this advantage creates a very difficult, unequal challenge for human opponents to overcome.

In the early fall of 2018, shortly after OpenAI's successful but incomplete accomplishments against professional *Dota 2* players, Chinese technology giant Tencent Holdings announced that it had developed a pair of virtual AI bots capable of playing *StarCraft II* not only better than humans, but also better than the most difficult "cheating" levels of the game itself—making it the first company and program capable of doing so. Tencent is a huge conglomerate that we'll discuss more in later chapters. It provides a variety of internet-related services and products, including entertainment, e-commerce, social media, and artificial intelligence.[36] It's also the world's largest gaming company and one of the largest venture capital firms and investment corporations in the world. Tencent controls hundreds of subsidiaries and associate corporations and, as with all Chinese state-owned corporations, its developments and intellectual property are generally available to the Chinese government and military. More about that later.

Tencent's pair of virtual AI bots designed to play *StarCraft II* were named TSTARBOT1 and TSTARBOT2. Like all neural network–based AI agents—which we'll describe in Chapter 9—both of the TSTARBOTs were designed to imitate human thinking, and they weren't given any advantages that humans don't have. They were even required to interface with the game through mouse clicks, albeit virtual, just as humans do. Further, they couldn't see anything that a human player wouldn't see. For them, the fog of war was intact, just as it is for any person or team of persons.

The first bot, TSTARBOT1, was coded with very specific algorithms designed to tackle lower-level game elements. The second and more robust bot, TSTARBOT2, was coded to handle the broader facets of the game. Thanks to that higher-level commander type of oversight, the duo was able to engage on the combative front line of play while still keeping track of overall developments and long-term strategies. In that way, the bots could operate in a more humanlike, hierarchical distribution of roles and responsibilities. As a result, Tencent's AI proved dominant, even against the cheating, more *knowledgeable* AI of *StarCraft II*. According to Tencent, when TSTARBOT2 was engaged to oversee game play, it won 90 percent of the time.

———

Regardless of whether an algorithm is playing a board game like Go, manipulating a video game like *StarCraft II*, or directing a self-driving car to operate safely within the traffic around it, everything AI applications are capable of doing depends upon their ability to obtain and analyze the necessary indicators of the situations they're tasked to solve. In essence, the quality of their performance depends upon the quality and completeness of the information available to them. With that in mind, it's time to look at the most fundamental form of that information. It's called data . . . and it constitutes the foundation upon which all learning algorithms depend—even the ones taking place between our own ears.

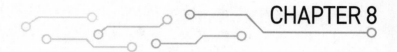

CHAPTER 8

A DELUGE OF DATA

It is a capital mistake to theorize before one has data. Insensibly, one begins to twist the facts to suit theories, instead of theories to suit facts.

—*Arthur Conan Doyle, 1859–1930*
Sherlock Holmes, "A Scandal in Bohemia" (1892)

Everything about our world is defined and characterized by data. It always has been. More and more, though, our lives are becoming described and influenced by the creation, preservation, and exchange of *digital* data. We hear about the explosion and importance of it on the news, at work, and in advertisements for products and services that touch all aspects of our lives. As the next chapter explains, digital data is the single resource that enables all of AI. To truly understand AI, it's therefore crucial to first understand the essence of data.

So, what is it? That might seem an easy question with an all-too-obvious answer, but it's not. Most people think of data as information that's already been sorted and structured to one extent or another—like the formatted and categorized information presented on a spreadsheet. But it's actually much more fundamental than that. Data is the most elementary form of information possible, before it has been sorted, categorized, compared, or analyzed in any manner. For everything that exists and for everything that happens in the physical and cultural world around us—including every activity, event, interaction, and thing (alive or otherwise)—the data that fundamentally defines and describes it, at some point, likewise exists.

Our ability to learn requires our capacity to acquire data and to analyze it. Without data, intelligence just isn't possible, not at any level—and not in any animal, individual, or computer. As humans, we observe and collect data about the world around us through our five natural senses. We see things, hear things, feel, smell, and taste things. As we make sensory observations about the world, our brains process and analyze the data we collect to interpret and understand what we've observed. In the course of that process, our brains turn the data of our observations into usable information by sorting and categorizing it through comparisons to what we've previously experienced and what we already know. Then, we analyze the information further, either in that moment or at a later point in time . . . and we learn. We learn *because* of the information, we learn *about* the information, and we learn *from* the information. In short, first we absorb data, then we process it into information, and then we learn from it to acquire knowledge—knowledge not only about the new information itself, but about everything else to which it may relate in our lives. Distinguishing in these ways between the words *data*, *information*, and *knowledge* might seem a confusing or trivial matter of semantics, but it's not. They are important distinctions to keep clear and to understand for a variety of matters that ultimately relate to the rudiments of AI.

———

With the entirety of the learning process dependent upon it, there are two fundamental truths about data. First, while the laws of quantum mechanics insist information about the past is never actually lost, for us humans, it most surely can be. Data, or certainly our ability to capture it, doesn't naturally last forever. It's fleeting. Unless someone or something is present to observe and somehow record the data emanating from an event when or as it occurs, then the data that describes the occurrence tends to disappear. Think of the tree that falls in the woods with no one there to observe it. Does it make a sound? Other than in the most esoteric of philosophic conversations, the answer is yes—for a falling tree certainly creates sound waves. But there's a slightly different and much more pertinent question to ask. Can we ever know exactly what it sounded like? To that question, the answer is no, not unless it was recorded in some way or unless some other evidence or effect of the sound exists that we can later still detect and measure—and from which we can recreate the sound. In other words, unless an event was observed when it occurred, or unless the event leaves a physical or measurable, residual trace we can later discover, the data of the event—as far as we humans are concerned—is lost.

Over the course of our time on the planet, we've managed to deal with the fleeting nature of data by creating a tremendous number of methods, instruments, and devices to help us observe data or detect its remnants, and then record our observations to make it usable to us, and to make it last. Today, we have technology that augments our capacity to see, hear, feel, and even to smell and taste. We've become incredibly adept at enhancing our ability to observe the details of things as they occur, just as we've become impressively capable of discovering the remaining evidence of things long after they've happened. We're always innovating new ways to observe and compile data. One wildly impressive example is our ability to detect radiation remaining from the earliest days of the universe. Called cosmic microwave background radiation (CMB), it is, literally, the detectable traces of heat left over from the time the energy of our universe first became visible just under 13.8 billion years ago.[1] A more recent example is the LIGO project's detection of gravitational waves traveling past our planet that originated from the violent convergence of black holes and neutron stars billions of light-years away.[2] And, even more extraordinary, is the first picture we ever compiled of an actual black hole. Located at the center of the M87 galaxy 55 million light-years away, it therefore took the faint photons (particles of light) surrounding the black hole 55 million years to reach our planet before a global array of telescopes detected them. Then through incredibly complex computer calculations, we were able to reconstruct a visual image of what we might see . . . if we were there.[3]

The second truth about data is that there is simply too much of it for us to fully perceive, collect, and manage on our own. The capacity of the human brain is enormous. But, as we've seen, there's a practical limit to the amount of data that any person or any group of people can ever obtain, let alone meaningfully process or share.

Our earliest computers promised the ability to eventually store amounts of data that far exceeded our own human abilities. Then, with programmable computers, we created the means of analyzing that data far faster than we alone ever could. Still, though, we had to specifically instruct our computers what to do. Because computers were only able to execute the tasks we instructed them to perform, we not only had to know what questions to ask, but also had to know—or at least speculate—what information might be relevant and what aspects of that information might be pertinent to the answers we were trying to obtain. Our earliest coding languages and algorithms therefore had to direct computers *what* data to analyze and *how* to analyze it. As a practical matter, we therefore had to theorize and hypothesize. We had to formulate the questions and anticipate what the answers might be in order to program our computers to perform the calculations and analyses necessary to find them. Even with computers to aid us, our own limitations therefore still stood as obstacles in our way.

———

To complicate matters even more, the amount of data, and all of the more structured information resulting from it, increased dramatically with the creation of the internet, an innovation that altered our world in profound ways.

In the early 1960s, an American computer scientist named Leonard Kleinrock recognized the benefits that could be achieved if computers were able to connect and share information between them. To make that possible, Kleinrock proposed a method of grouping data into digital packets so they could be transmitted over a network of interconnected computers, allowing users to exchange and share volumes of information in a way never before possible.[4] Kleinrock's concept quickly led to the development of *packet switching* and to other processes that enabled the US Defense Department's Advanced Research Projects Agency (DARPA) to develop the very first computer-to-computer link. Called ARPANET, this first network initially connected four computers located at the University of California, Los Angeles (UCLA); the University of California, Santa Barbara (UCSB); Stanford Research Institute (SRI); and the University of Utah. On October 29, 1969, the very first computer message was sent from UCLA to SRI and, with it, the internet (which is an abbreviation for "interconnected network") was born.[5] Throughout the 1970s and '80s, other organizations like the National Science Foundation (NSF), the Computer Science Network (CSNET), and various other universities and research institutions continued to expand networking capabilities and interconnectivity as the internet gradually evolved.[6]

At first, the growth of the internet was tediously slow. Until the early 1990s, the network wasn't available to the general public. And even for those with access, most of whom were academics and researchers, sharing information was cumbersome and difficult to accomplish—primarily because different computers that were owned and operated by different institutions tended to use different programming languages. Therefore, even when computers were connected, the operators of those networked computers still had to find a translatable common ground between them. People operating one computer had to know, or learn, the programming code of the computer from which they wanted to gain information. It was the recurring problem of different people speaking different languages.

In 1989, an English software engineer named Tim Berners-Lee proposed a solution. While working at the European Organization for Nuclear Research (CERN), Berners-Lee wrote a paper called "Information Management: A Proposal," in which he suggested that connected computers could share information using a newly emerging technology called hypertext—which is text displayed on a computer or other device in hyperlinks that, with a simple click, immediately connect the user to other text, documents, and internet locations.[7] Hypertext is so fundamentally common to us now that we don't even think of it. But when first introduced, it changed everything about the practical functionality of the internet.

Berners-Lee called his new process of network connectivity the World Wide Web, and he launched the very first web page in August 1991.[8] It wasn't long before the web became mainstream. By the mid-1990s, it was commonly available through various public browsers (the precursors of Google Chrome, Microsoft Edge, Mozilla Firefox, and Apple Safari that we know today) that suddenly gave internet users the ability to search for information by topic and to immediately link to web pages that contain it—without needing to know any interpretive programming or coding themselves.

———

The amount of data that we create and now share via the internet is enormous. With just under eight billion people in the world, more than half of us, over four billion, now use the internet on a regular basis. And most of us use multiple devices to do so—desktops, laptops, tablets, mobile phones, televisions, cars, watches, and even home appliances. By 2025, it's estimated there will be more than 75 billion active smart devices around the world—which is almost 10 devices for every person on the planet, making for a never-ending and ever-increasing process of digital data generation.[9]

In fact, more than 90 percent of all data available in the world has been generated in just the last few years alone.[10] Although that sounds staggering, it's not so surprising given the numbers of people that are digitally engaged across the globe, the numbers of ways we're now able to digitally connect, and the permanence of the digital data we create with each connection. Every time we engage with the internet—whether to interact on social media, to entertain ourselves, to manage our personal affairs, to search for information, to make purchases, or to just connect and communicate with each other via email—we flood the internet with data about our interests, beliefs, and behaviors. And all of it now survives, arguably forever.

With all of this connectivity, we now create 2.5 trillion bytes of new data each day. Two billion people are active on Facebook alone, 1.5 billion of them every day. We conduct 3.5 billion Google searches each day. We upload 300 million photos to various internet-based platforms daily. And, every *minute* of every day, we send 16 million text messages, send 156 million emails, and make 155,000 Skype calls.[11]

There's very little we can't do on the internet. From rocks to rockets, we can purchase most anything. There are now more than 1.5 billion websites—nearly one for every five people on the planet—that connect us to just about any information, service, or product we could possibly want.[12] Amazon, the largest of the commercial platforms, was initially launched in 1995 as just an online bookstore. It currently offers more than 500 million products of all types to US internet users and more than 3 billion different products to users worldwide. Catering to markets primarily in America and Europe, on a busy day Amazon now

World Population by Country			
Rank	Country	Population (2019)	World Share
1	China	1,420,062,022	18.41%
2	India	1,368,737,513	17.74%
3	United States	329,093,110	4.27%
4	Indonesia	269,536,482	3.49%
5	Brazil	212,392,717	2.75%
6	Pakistan	204,596,442	2.65%
7	Nigeria	200,962,417	2.60%
8	Bangladesh	168,065,920	2.18%
9	Russia	143,895,551	1.87%
10	Mexico	132,328,035	1.72%

Figure 8.1: Ten most populated countries, 2019.

sells upward of 400 products per second, accounting for 49 percent of all online retail sales in the US, and more than 5 percent of all retail sales overall.[13] Amazon's Chinese counterpart, an e-commerce company called Alibaba, which we'll discuss more in Chapter 13, has an even more impressive share of the Asian market, accounting for approximately 58 percent of all Chinese online retail sales.[14]

———

Digital data is now commonly referred to as "the new oil." Although the phrase has become cliché, it's essentially accurate. Even before data is structured or manipulated in any way, the variety and amounts of it that we generate—about ourselves, our families, our communities, and even our cultures—are of immense and powerful commercial and political value. But, just like oil, it's also spread unevenly across geographies and nations.

China is the most populated country on Earth, with more than 1.4 billion people. India is a close second with over 1.35 billion, and the US is in a very distant third place with a total population of fewer than 330

World Population by Region

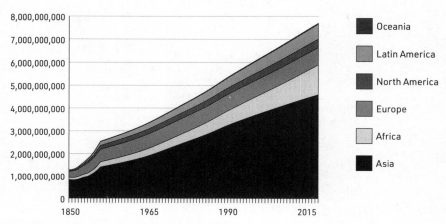

Figure 8.2: World population growth, by region.

million people—which is less than a quarter of the population of either China or India, and only 4.4 percent of the world's total population. See Figure 8.1. Regionally, Asia represents 63 percent of the globe's populace, Africa 16 percent, Latin America 9 percent, and Europe 7.5 percent.[15] See Figure 8.2.

An impressive 90 percent of the US population is connected to the internet, which amounts to about 300 million people. By contrast, less than 60 percent of China's population is currently digitally connected, but even that smaller percentage of its population nonetheless equates to more than 800 million people, almost three times the number of Americans. In fact, of all people using the internet globally, 49 percent of them are Chinese and Southeast Asian, 17 percent are European, 11 percent are African, and just over 10 percent are Latin American. Only 8 percent of all internet traffic originates from the US and, because of the rising number of users from other countries, the US percentage will only continue to decrease.[16] See Figure 8.3. The value of the disproportionate amounts of data created by these populations, as it relates to AI and particularly as it relates to China, will become apparent in later chapters.

Internet Users by World Region

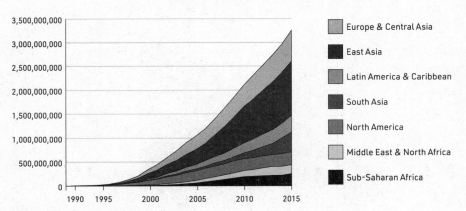

Figure 8.3: Global internet usage and growth, by region.

With all of the data, programs, applications, and computing power that became increasingly available after the creation of the internet and World Wide Web, it wasn't long before it became apparent that traditional approaches to computer storage just wouldn't be sufficient. Historically, everything that was needed to perform our computing processes was stored on the hard drives of our individual computers or within the combined memory of our local, self-contained networks. But, as technology and data volumes continued to expand, we were forced to adapt. The costs, time, and human energy required to continue buying and updating our hardware, software, memory, and security quickly became unmanageable. Our investments and capabilities became obsolete almost as soon as we unboxed and learned how to use our latest devices and the software that supported them. The situation was expensive, inefficient, and often less than secure.

One solution that seemed to make sense for individual and business consumers alike had been futuristically theorized, at least in concept, for decades. In 1996, a small group of Compaq Computer executives proposed a theoretical structure for the concept and gave it a name. In a document distributed only internally at the company's offices outside of Houston, they plotted the future of the internet and imagined a day when software and storage capacity would be disbursed and shared throughout the web. They called it "cloud computing."[17]

By the early 2000s, the concept of the cloud was taking shape in the real world. Contrary to what many people think, there's nothing mysterious about it. Cloud computing is just a means of accessing all aspects of computing services *over* the internet—including servers, storage, databases, software, security, and even artificial intelligence. In practice, the cloud is nothing more than a network, albeit vast, that allows you to utilize other computers' hardware and software.

Most of us use cloud computing all the time without even realizing it. When we use our laptops, iPhones, or any other devices to type a Google search query, our personal device doesn't really have much to do with finding the information we're looking for. It's only acting as a messenger that communicates our search terms to an array of other computers somewhere out in the world. Those computers then use their own programs and databases to determine the results and send them back to us. It all occurs at incredible speed, and, for all we know, the real work that identified the information we requested may have been performed by computers five miles away or on the other side of the planet. Regardless of their locations, cloud computing simply allows us to instantaneously draw upon the strength, software, and information of other computers—which are inevitably more powerful, equipped, and versatile than our own.

In addition to Google Search, other examples of cloud services that we frequently use include web-based email and cloud backup services for our phones and other computer devices. Even Netflix uses cloud computing to facilitate their video streaming service, and the cloud has likewise become the primary delivery method for the majority of apps now available, particularly from companies that offer their applications free of charge or for subscription fees over the internet, rather than as stand-alone products that require full downloads.

The infrastructures required for cloud services are provided primarily by a handful of major commercial cloud providers. The largest two are Microsoft Azure and Amazon Web Services (AWS)—the latter of which launched the first public cloud service in 2006 as a way of turning its unused computer power into commercial revenue. Each of those providers now generates close to $30 billion per year from their cloud services.[18] IBM, Oracle, and Google are the other main American companies offering cloud computing, and Alibaba is the principal Chinese provider of cloud services. These companies manage networks of huge, secure data centers that are usually spread over broad geographic regions where they house the infrastructures that power and store the data, systems, and software necessary to operate their clouds. By spreading out the data centers geographically, and by also breaking each geographic region into smaller zones of supporting facilities, they can keep their customers' data somewhat localized while still protecting it against local calamities, power outages, and other events—either natural or man-made—that might compromise the facilities and the information contained within them.

Cloud operations can either be public, private, or hybrid. Public clouds allow all users to share space and time on the cloud and to access it through unrestricted means. Public is the usual cloud format for individual and personal cloud computing, but many companies also opt to use public cloud infrastructures for their internet email systems and for employees who share documents using Google Drive.

Private clouds work technologically the same as their public counterparts, except they service a single company and require authorized access to use the network. They can be managed either exclusively by the user company or by one of the major cloud providers on the company's behalf. Either way, a private cloud is usually fully integrated with the company's existing infrastructure, network of users, and databases . . . and can span countries and continents just as a public cloud can.

Oftentimes, companies have needs that lie somewhere between public and private clouds, so they opt instead for a hybrid cloud, which, just as the name implies, provides elements of each that usually reflect the different levels of security and corporate control required for various cloud-based purposes and activities.

Security has always been a question and an area of significant concern when it comes to cloud computing. People are understandably curious about how their data can remain safe when traveling back and forth over the internet. In reality, though, cloud-based systems are generally far more secure than conventional, localized ones. There's an economy of scale and expertise that applies and that makes providers of cloud-based services much more capable of effective security measures than smaller, independent users. Large, sophisticated providers like Amazon, Microsoft, and Google are far more capable of providing qualified IT personnel, cutting-edge security programs, and carefully constructed and secure facilities than individual citizens or most companies could ever hope to acquire or properly manage themselves. As a result, security is actually a compelling reason to use cloud-based systems rather than avoid them.

Privacy is a more nuanced and complex issue than security. While we all understand what we mean by keeping data secure from outside corruption or invasion, it's a little less clear what we can, or should, expect when it comes to keeping our data private from the cloud providers themselves. One of the most significant concerns has to do with what's referred to as big data—which is the statistical, analytic type of information the cloud providers can gather, overall, from all of the ways the masses use their cloud-based services—like what websites we visit, what entertainment or social preferences we have, what products we purchase, what news we read, what political leanings we exhibit, and so on. In reality, there is a quid pro quo for using "free" web-based services and apps like Google, Facebook, Twitter, Instagram, Snapchat, and

TikTok. There's a completely reasonable argument that we do in fact *pay* for what we use by giving those who provide such services consumer and behavioral information about us that they can then use for their own commercial benefits—to generate revenue, for instance, through targeted advertising campaigns. As can rightly be said, "If you don't think you're paying for it, that just means you're not the customer—you're the product."

Another complication with our privacy expectations is that privacy rights are governed differently in different parts of the world, some of which are more lenient and others more protective. For instance, the European Union (EU, which we'll discuss more in later chapters) strictly regulates how data can be transferred from one country to another, and even how companies like Google that have multiple subsidiaries operating across different countries and continents can transfer or share corporately owned data. Although internet-based cloud computing makes national boundaries somewhat obsolete, national and international laws do still operate according to previous standards. At best, though, those laws are often difficult to enforce when it comes to digital data. After all, this new "oil" doesn't require a conspicuous barrel, shipping container, or cargo hold to transport it—and tracking its distribution can therefore be a tremendously difficult task.

CHAPTER 9

MIMICKING THE MIND

If you wish to make an apple pie from scratch, you must first invent the universe.

—Carl Sagan, 1934–1996
Astronomer, Astrophysicist, Author
Cosmos, TV Adaptation

F or those in the trenches of AI research, development, and implementation, the technical aspects are complex and require significant expertise. That's why the field is driven by the best and brightest minds available. But, for the rest of us, there's good news. We don't need to initiate the next great breakthroughs ourselves, nor do we need to design hardware or write algorithms. All *we* need is to generally understand how and why they work. Fortunately, an accurate and conversationally sufficient explanation of the key processes behind today's AI doesn't have to be complicated or hopelessly confusing.

The single, underlying goal of AI is to empower computer systems to perform the higher kinds of intellectual functions we've traditionally thought are only possible by humans. That sounds like a pretty broad and ambiguous statement, but what it really means is this: the aim of AI is to create computer systems that learn to perform better as they acquire more data and objective experiences related to whatever task they're designed to accomplish. That's what humans do. And with that single goal in mind, the question becomes simple. How do you design a computer that learns, that develops *knowledge*, from data?

The answer that's directing the current path of AI resulted from looking at the structures and processes of our own minds for guidance. Through our five natural senses, humans continually acquire information that our brains evaluate through one of two general methods. The first of those is symbol based. Through vision, we analyze data primarily in a symbolic manner—evaluating the constant flow of incoming images by comparing them to what we've seen before so we can identify the objects and actions—the information—that they represent. In other words, from a biological perspective, images are symbolic of the information they contain. There's a lot of truth to the saying that a picture paints a thousand words. Conversely, when we obtain information through language—whether verbal or written—we learn in more of a data-based, statistical manner. It's the individual words, and the infinite number of ways they can be combined, that accumulate to inform us of overall meanings, nuances, and details.

Following World War II, and until the late 1980s, the main thrust of AI research concentrated on a symbolic-based approach. In these systems, also known in the industry as Good, Old-Fashioned AI (GOFAI), the human programmers themselves had to first understand the relationships existing within the data that they'd eventually ask a computer to analyze.[1] Then the programmers had to translate their understanding of those relationships into their computer-coded algorithms. Essentially, they had to instruct the computer, "if this, then that," at every step of the way. While those systems could be created to function efficiently and accurately, they were also unavoidably static, as they always applied the same, human-provided rules to the data they were fed. There wasn't much new "learning" going on, at least not by the computer. This approach to AI is also called "symbolic, *knowledge-based* learning" or "*expert* learning," because it foundationally depends upon the knowledge that human experts already possess.

While proponents of the symbolic reasoning approach to AI were pursuing their course of research and development throughout the 1970s, '80s, and '90s, others were pursuing a distinctly different method, one that was more oriented toward a statistical analysis of data. Led primarily by cognitive psychologist and computer scientist Geoffrey Hinton, who is now with the University of Toronto and Google, their theory wasn't focused upon applying hard-coded instructions *to* data, but instead upon the correlations the algorithms themselves could discern *from* data.[2] As far back as 1959, the term "machine learning" was used to refer to a proposed AI system based upon a hardware and software design that simulated the neurons and synaptic connections of the human brain—which we discussed in Chapter 6. The design is called an *artificial neural network*, and the idea was that the system would be able to make sense of data itself, without any direct instructions or guiding knowledge from humans.

But Hinton and other early advocates of the idea faced two seemingly insurmountable problems. To function effectively, it was clear that an artificial neural network would require two things: huge amounts of computing power and enormous amounts of data. Neither of those was readily available at the time. And while scientists accurately anticipated that the required amount of computing power would eventually become available, most data scientists, even during the 1980s and '90s, couldn't foresee a day when the required amount of data would realistically be available—or efficiently obtainable. The world's population still conducted itself almost exclusively in analog ways . . . and there just wasn't that much digital data to be found.

As a result, the machine learning/neural network approach to AI didn't attract a lot of mainstream confidence or support. But then, beginning in the late 1990s, the field began looking more viable. Just as Moore's Law had predicted, computing power was continuing to grow exponentially. Likewise, through the spread of the internet and the outbreak of online activities we discussed earlier, the amount of digitized data the world was creating suddenly exploded. Better yet, a lot of that data was suddenly obtainable through the very thing facilitating its creation, the World Wide Web.

While those changes were occurring, Hinton and like-minded researchers had continued their dogged focus. In 2012, machine learning finally proved effective when a major breakthrough in computer vision capabilities occurred that changed the attitudes of most naysayers.[3] A new capability called *deep machine learning* opened the doors to AI as we know it today, and Hinton and his colleagues were behind it. In fact, in March 2019, he and two of his principal research colleagues—Yoshua Bengio at the University of Montreal and Yann LeCun at the Courant Institute of Mathematical Sciences at New York University—were eventually awarded the A. M. Turing Award in recognition of their pioneering work.[4] Within the world of computer science, the award carries the prestige of a Nobel Prize.

Machine learning is the approach to AI that's most useful when it isn't practical, possible, or preferential to create algorithms that specifically instruct a computer *how* to perform a given task. When we need or want a computer to sort out the strategies and answers for itself, machine learning is the solution. Imagine all of the specific programming directions that would have been required for DeepMind's AlphaGo program to beat Lee Sedol in their 2016 Go match. More importantly, try to imagine how the DeepMind team could even have known what those instructions should be, when, at the end of it all, AlphaGo itself proved that our very best players had never even conceived of its winning strategies and moves. Likewise, how many different instructions would a computer need in order to navigate the best street route, particularly if we want it to evaluate endless traffic variables, road conditions, and other factors that might potentially be relevant? And how could we possibly program a computer to consider all of the elements relevant for selecting a promising new song we might like (yes, algorithms that make playlist recommendations are machine learning too), when we're not even consciously aware of all the subjective elements ourselves? The point is this: there are patterns to the world, and to ourselves, that we just can't always discern . . . and that we therefore can't possibly articulate to ourselves, let alone to a computer.

Machine learning systems are therefore designed to perform without having to rely upon human instructions that tell them exactly what to do or how to accomplish their evaluative tasks. Instead, the algorithms instruct—or more aptly, empower—the system to search for patterns and factual truths by discerning values of importance, or weights of significance, from large sets of data that are fed into them, and from which they can determine statistical conclusions or draw increasingly more accurate probabilities. For a computer, data is the equivalent of experience. So, the more data that a machine learning system processes, the better it becomes.

———

As mentioned earlier, an artificial neural network is modeled on the network of biological neurons in a brain. The technique is used to build machine learning algorithms that can process data in a layered approach, weigh the relative value or significance of the incoming data, and then transmit a measurement of that significance to all related nodes at the next layer. This forward-feeding mathematical process allows the system to compute, at each layer, the relative importance of all the various factors the data set contains and then, at the end of the layering chain, to output the results of its combined, systematic calculations.

Fundamentally, the purpose of these networks is to distinguish the specific characteristics of individual pieces of data and to identify and measure similarities and differences within the context of the overall set—essentially, to find patterns. It's extremely complicated technology that requires precise, finely tuned design parameters. In many ways, though, it's arguably no different than a hypothetical stack of spreadsheets . . . all of which are interconnected and working together to sort out the defining characteristics of a common set of data. At the start of the process, the data is raw and the relative significance of its characteristics are unknown. But, by the end of the process, distinguishing qualities and patterns emerge from which the system can categorize and classify the data, and even make predictions.

In the earliest days, neural networks were *shallow*, usually consisting of only a few layers: an input layer, a middle or *hidden* layer, and an output layer. Data is fed into the input layer, analyzed, and weighed as it progresses through the hidden layer, and it's then forwarded to the output layer as a measured result. Nowadays, the process works generally the same, but the frameworks of the networks often include many middle, hidden layers, sometimes thousands. These are called *deep* neural networks, or deep-learning systems. See Figure 9.1. With the addition of each new layer in the network, the system is better able to discern specific and deeper levels of minutiae about the data passing through it. And to make the networks capable of even more complex analysis, the middle layers are often designed to act differently from one another—some might pass all data through them to the next layers, some might combine data before forwarding it on, and some might dispose of data entirely if it's deemed irrelevant. Through a process called *back propagation*, the results of any measurements—at any points in the process—can even be fed back to prior layers over and over again to continually adjust the weights and measurements based on the overall dynamics of the evolving analysis.

We refer to the middle layers as hidden because we don't know the exact measurements the system is assigning to the data at any one of the interim points in the process. Keeping in mind that all of this takes place

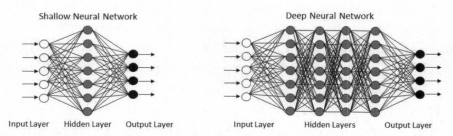

Figure 9.1: Shallow and deep neural networks. Data is fed into the input layer, analyzed, and weighed as it progresses through the hidden layer(s), and is then forwarded to the output layer.

very quickly, in nanoseconds, we can't watch each of the instantaneous interconnections and interactions occurring in the hidden layers. But that doesn't mean there's some mystical or inherently unreliable process playing out within them. It only means that we see the end result of the assessment, not the interim steps, just as we do in the case of our own minds. If a neural net could describe "how" or "why" it came to the conclusion it did, the description wouldn't be much different from what we would say about our own assessments of information streaming into our own brains. We would probably say something like, "Well, I *thought* about it and just concluded what I did—based, I imagine, on all of my prior experiences and the information that was newly available to me." Similarly, a machine learning system would say that it assessed the information, made its calculations, and *predicted* that its output was accurate, or most probable. In both cases, although we might have no idea of the precise measurements occurring deep at the evaluative levels, at the end of the process we feel confident that we've made sense of the data, and that we can likewise feel confident of whatever we've concluded.

———

Machine learning algorithms tend to perform better with every exposure to new data, much like we get better at things with each new experience.

First, though, algorithms must learn to accomplish their purpose. They must be trained. To do so, they're initially fed sets of data appropriately called *training data*. In the next chapter, we'll discuss some of the latent problems that can occur depending upon unintended attributes of the training data, like inherent biases, but for now, it's just important to know that it's at this stage of development when machine learning algorithms are tested, fine-tuned, and first able to prove their abilities.

Regardless of whether a machine learning program is fed data composed of images, numbers, or natural language at the training stage, there are three main types of machine learning systems—supervised, unsupervised, and reinforcement.

In supervised learning systems, the training data is first labeled by humans with the correct classification or output value. For example, training images fed into a system designed to identify certain types of fruit, through computer vision, would already have a "banana" output label for all of the images showing that particular fruit. Likewise, training data fed into a system intended to eventually identify diabetes sufferers through their statistical medical records would include sample records that already have a "diabetes" output label when they reflect a medical history that gave rise to that condition. From there, both machine learning systems would search to find the distinguishing characteristics—and patterns of characteristics—in the data that result in the banana and diabetes labels. Afterward, having learned what characteristics to look for, the systems can be used to process new sets of data, without the labels, to identify new, unlabeled images of bananas or, in the other system's case, people statistically likely to develop diabetes.

For unsupervised machine learning algorithms, the training data isn't classified or labeled in any manner before being fed into the system. Instead, the system analyzes the data without any prior guidance or specific goal. Its task is more generally to discover, on its own, any similarities or distinct and recurring differences within the data so it can be grouped or consolidated according to those qualities. In these systems, the machine learning application is essentially being asked to find unifying or distinguishing characteristics within the data from which categories can *then* be determined and labeled. This kind of approach is frequently used to explore data in order to find hidden commonalities within very broad sets of complex and varied data. It's often referred to as *cluster analysis*, and is routinely used for market research—where, for advertising and other strategies, it's extremely valuable to find common characteristics, behaviors, and habits within otherwise broad bands of prospective consumers whose similarities aren't readily apparent.

Reinforcement learning is similar to unsupervised learning in that the training data isn't labeled. But when the system draws a conclusion about the data or acts upon it in some way, the outcome is graded, or rewarded, and the algorithm accordingly learns what actions are most valued. For this reason, and as we learned in Chapter 7, reinforcement learning is particularly well suited for game play and other goal-oriented, competitive tasks. When a reinforcement algorithm plays a game—which is nothing more than an incoming stream of data that it's tasked to assess and act upon—and accomplishes a desired goal or a win, then a reward is obtained, *points* for instance. As a result, the system learns that the moves or steps that caused that outcome were strategically effective, so it "reinforces" them as having been good moves. This isn't overly effective if the computer plays only a small number of games, since strong moves in one situation may not be strong in others. But if it plays thousands or millions of games, which can be done in just moments or hours at today's computer speeds, then the cumulative effect of the reinforcement method is extremely efficient for determining winning strategies.

———

While the type of training data from which a neural network is designed to learn defines whether it's a supervised, unsupervised, or reinforcement learning system, there are a few other classifications of neural networks important to understand as well. They are convolutional neural networks (CNNs), recurrent neural networks (RNNs), and generative adversarial networks (GANs). Each of them operates differently from the others, and each is particularly well suited to different kinds of applications.

Convolutional neural networks are the most commonly used network for computer vision programs or any machine learning applications that require the system to recognize images or shapes. CNNs operate in a way that roughly resembles the processes of our own visual cortex. They're very complex, multilayered program structures that generally rely upon powerful, parallel computing hardware to make them as dimensionally perceptive as possible. An easy way to understand them is to simply imagine that each of the artificial neurons, or nodes, in the layers of the CNN are set to identify the color, shape, or texture of a specific point, or pixel, on an image and to assign to it a specific numeric value. That value, or measurement, is then passed on to the next layer of the neural network. As this process repeats over and over again, the image is essentially scanned from left to right, and from top to bottom, so the output layer can measure the most discernable characteristics of the image to identify the most probable object or scene that it depicts. CNNs are routinely used to recognize things like faces, animals, stop signs, cars, trees, and even text (since the shapes of letters are their most defining and recognizable characteristics). CNNs usually utilize a supervised approach to learning and are generally trained by many thousands or millions of images that humans previously labeled and fed into them, thereby giving the CNN systems the ability to later recognize new objects that are either identical or similar to those they've previously learned to identify. This is how a self-driving car can identify a yield sign, even when it's damaged, faded, or partially covered by snow.

While CNNs work well for image recognition and object classification, recurrent neural networks work well for sequential data that requires interpretive nuance and a broad context in order to fully understand the input data. RNNs are commonly used for voice recognition programs and natural language processing programs (NLPs) that

interpret human speech and written language. Whereas CNNs only utilize the input data, RNNs are able to momentarily save the output of
a layer and feed it back to the input in order to better predict the next
likely output. It's as though RNNs have internal, short-term memories
that allow them not only to identify single data points, like an image, but
also to smoothly transition entire sequences of data, videos, speech, and
written narratives.

Finally, generative adversarial networks, GANs, are composed
of two separate, deep neural networks designed to work against one
another. One of the neural networks, called the *generator*, creates new
data that the other network, called the *discriminator*, evaluates in order
to determine if is indistinguishable from other data in the training set,
in which case it is considered *authentic*. Essentially, the generator is trying to create information—like a paragraph of text, an image, or even a
voice rendition—that the discriminator is unable to identify as fake. As
the discriminator rejects unsuccessful attempts to fool it, the generator
gets progressively better, learning from its failed attempts and refining
its skills. GANs are most often used to create new data that's meant
to mimic or simulate what's naturally created or crafted by humans—
like prose, paintings, images, videos, speech, and music. They're often
called bot artists since they can be used to add the kind of content to our
environment that we once believed was only possible through human
creativity. As you're probably imagining, and as we'll address more in
Chapters 11 and 14, they can be used quite effectively for disinformation
campaigns, deepfake videos, and other manipulative, nefarious purposes.

———

The current and immediately foreseeable uses of machine learning applications are practically limitless. They can help us more effectively and
efficiently analyze, learn, and accomplish tasks that are already an integral part of our lives. They can also enable us to do entirely new things
that we'd never be able to consider or attempt without them. Whether we

perceive that as good or bad, exciting or distressing, it's our new reality. Machine learning is here to stay, and its uses are expanding every day. The following is a list of tasks at which machine learning applications have already proven well suited. Despite its length, the list is far from complete.

Aerospace and aeronautics—research and design

Agriculture—crop management and control

Authenticity verification—visual, voice, and data

Aviation—logistics, routing, and traffic control

Biometrics—assessment and prescriptives

Climate—analysis and prediction

Computer hardware design and engineering

Computer vision

Counterterrorism

Crime analysis and detection

Customer relations—proactive management

Customer service—call centers and response

Cybersecurity

Data analysis—for any use

Disaster response, recovery, and resupply

Disease detection and contact tracing

Disinformation and deepfake detection

DNA sequencing and classification

Due diligence research

Economics—analysis and prediction

Education—curricula, content, and proficiencies

Emergency detection and response

Energy—control, creation, efficiency, and optimization

Entertainment—preferences, creation, and delivery

Environmental impact and conservation analyses

Finances—personal, business

Financial services—market analysis, trading

Food—processing, preservation, distribution

Forest fire prediction, control, and containment

Fraud detection—online, identity, credit

Game creation and competitive play

Handwriting recognition

Harvesting—agriculture

Healthcare—plan management and diagnostics

Image processing

Information retrieval and data mining

Insurance underwriting

Internet fraud detection

Language translation—verbal and written

Law enforcement

Legal—research, analysis, and writing

Logistics—supply and distribution

Manufacturing—facilities and processes

Market analysis

Marketing strategies

Media—customer preferences and content

Medical—research, diagnosis, and treatment

Military—all aspects, like any other enterprise

National security

Natural language processing

Navigation—land, air, and water

News verification—authenticity and fact checking

Online advertising

Performance analysis—materials, products, people

Personnel—assessment and optimization

Pharmaceutical—research and development

Politics—analysis, polling, and messaging

Products—design, manufacturing, and assembly

Proving mathematical theorems

Quality control—products and services

Retail—inventory and pricing

Robotics—spatial assessment and locomotion

Scientific research—all branches

Search engine optimization

Security—premises, personal, and virtual

Self-driving, self-flying, self-sailing vehicles

Shipping—logistics, sorting, and handling

Social media—networking, implementation, analysis

Software design and engineering

Space exploration

Speech recognition

Telecommunications—service and efficiencies

Transportation—all types, all facets

Vaccines—research, development, and delivery

Weather—analysis and prediction

Wildlife research, assessment, and conservation

———

As the list makes clear, machine learning applications are already extremely good at performing tasks related to most areas of our lives. All of those tasks, however, are single, specialized, and specific. Some may seem broad, but they all function within a narrow domain of operation. Even the broadest of them, like computer vision, speech recognition, and natural language processing, are still at a limited, narrowly purposed stage. As sophisticated as these applications are, and as impressive as their performance has become, they're nonetheless only capable of accomplishing the particular task for which they were designed. Unlike humans, machine learning applications are not capable of applying strategies, knowledge, or skills acquired in one area to another. They're therefore called *narrow AI*. And, regardless of how things may seem to be progressing, with current technology and technologically viable approaches, the same algorithm that evaluates our investments just won't be able to also turn on a light when a cloud passes by our window, or even order a pizza for us when we're hungry. Narrow AI is therefore also referred to as *weak AI*. That doesn't mean, however, that it's inefficient or incapable. It's not. Narrow AI is very strong, efficient, and quite capable at its purposed job. It's just incompetent at anything beyond it.

When usable applications of AI were first emerging in the early 2010s, many people, AI experts included, spent a great deal of time and conversation speculating about the theoretical extremes AI might become capable of reaching at some point in the future. Those conversations continue, with people still commonly discussing the implications mankind might face if AI ever achieves an ability to perform outside of its design domain and beyond its intended task—if it ever becomes capable of broad, *general* intelligence or, worse, if it ever becomes capable of what's come to be called *superintelligence*—an imaginative technology capable of exceeding humans in all areas of knowledge at once. It's not my intent to argue that those conversations don't have value. They do, in certain contexts. But it's important to understand that neither artificial general intelligence (AGI) nor superintelligence is within the reach or even the possible range of *current* AI design, or any currently known technology.

AGI, also known as strong AI, is the type of hypothetical artificial intelligence that could operate beyond a single domain of information or task orientation, and that could perform successfully at any intellectual task just as well as a human. While we've been hypothesizing that AGI is right around the corner for many decades, the more we've accomplished in the science of AI, and the more we've come to appreciate the deepest mysteries of the human brain, the more we've realized just how hard it would be for a machine to become capable of achieving anything other than specifically oriented tasks. Humans may not be able to process data as fast or as comprehensively as computers, but we can think abstractly and across purposes. We can plan and sometimes even intuit the solutions to problems, at a general level, without even analyzing the details. This is what we sometimes characterize as *common sense*, which is something computers simply don't have—and that no currently known technology can give them. For the foreseeable future, AI will not be able to create or solve anything from something that isn't there, from data it doesn't specifically have, or for something it hasn't specifically been designed and trained to do. AI has no intuitive or transferrable abilities—and, for now, it's not going to acquire those abilities. Machine learning technology just doesn't allow for it. That's not to say, unequivocally, that general AI will never occur. But it would require a new breakthrough and an entirely different technological approach than those described in the previous pages. Whether such a breakthrough occurs at some point in the future remains, like all things, to be seen. But it's not on the visible horizon.

Beyond the presently hypothetical realm of general AI is an even more speculative concept: superintelligence—an imaginative technology capable of exceeding humans in all areas of knowledge at once. It's an unnerving concept to be sure, and one that can easily alarm our imagination and excite our fears. But, again, it would require something—and we're not sure what—even more technologically distant and currently unimaginable than whatever breakthrough might allow for general AI. Swedish philosopher and author Nick Bostrom is most notably associated with the concept. In his 2014 bestselling book, *Superintelligence: Paths, Dangers, Strategies*,[5] Bostrom warned of a theoretically possible future state of AI when it might become so much smarter than us that it could radically outperform even the most intelligent, knowledgeable, and expert human in every field and at every endeavor—generally, mathematically, scientifically, and socially—and be able to analyze situations and discern solutions to any problem on any topic. The point at which a machine might cross into that hypothetical state is called *the singularity*—a moment when all knowledge a superintelligent machine possesses runs exponentially rampant, with knowledge building upon knowledge in an endlessly escalating, unstoppable process that would also prompt the machine to divine its own motivations and purposes, arguably its own iteration of consciousness. It's fascinating to speculate about what would happen if that moment ever occurred. But whether it would be the best or worst thing that ever happened to the human race, or something in between, is only a hypothetical or philosophical question for now, as it's not a practically possible outcome of any known science.

In his book, Bostrom focused on the controls we might want to consider in order to avoid the many risks he presented that could result from a superintelligent AI. He called it the control problem. After the book's original publication, which was shortly before the major advances and practical implementations of machine learning we've discussed, many people seemed to ignore the fact that Bostrom himself made it clear that even in the worst scenario, superintelligence is an indeterminable length of time away. Discussion and commentaries inspired by the book tended to focus on the ultimate fear factors rather than the practical control problems or constraints we should pursue or at least philosophically consider. At the time, though, those fear-focused conversations were understandable . . . because there was an available space—because there was time and opportunity for speculative conversation.

Now, though, with what's *actually* transpiring in deep machine learning applications and capabilities, it has become abundantly clear that narrow AI presents enough actual and immediate concerns to warrant our full attention. Neither AGI nor superintelligence should deflect our focus or misdirect our attention. There are real matters to address, right now and right in front of us. When the fire is at the front door, there seems to me no reason to speculate about the lightning that might strike in the universe of all future possibilities. As you'll see a bit later in Part 3, there are alarming purposes behind narrow AI applications already causing real and detrimental effects in various parts of the world.

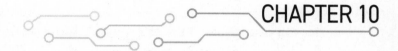

CHAPTER 10

BIAS IN THE MACHINE

On two occasions, I have been asked by members of Parliament, "Pray, Mr. Babbage, if you put into the machine wrong figures, will the right answers come out?" I am not able to rightly apprehend the kind of confusion of ideas that could provoke such a question.

—Charles Babbage, 1791–1871
English Mathematician, Philosopher, Engineer, and Inventor

Machine learning applications are designed to analyze data and formulate predictions without any overall guidance from us. That doesn't mean, however, that machine learning is necessarily safe from the effects or influence of our human biases. It's not. Just because an algorithm's analysis is based only on data doesn't mean its output will be neutral or objectively fair. It's quite common for human biases to be reflected in our data and, when they are, it stands to reason that any analyses, strategies, or predictions based on that data will be biased as well. Worse, if decisions are made or actions are taken based on biased analyses, then the underlying biases will of course perpetuate, and possibly ingrain, historical or cultural inequities even deeper into our lives.

The steps necessary to ensure that doesn't happen are difficult to accomplish—but not impossible. They require conscientious and concerted efforts at the development and training stages of machine learning algorithms, plus attentive analysis and oversight at the implementation and use stages. The first and most important step is to understand and acknowledge the underlying nature of the problem.

First, let's talk about human biases. Most of us believe that we're fully aware and consciously in control of our biased inclinations and opinions, and that we're able to intentionally include or exclude them however we see fit during the never-ending course of decisions we make throughout our days. But, we're not. The truth is, we're relatively unable to separate ourselves from our biases, or our biases from ourselves. We aren't even aware of many prejudices we hold, and we're accordingly unaware of the many ways they influence our behavior. Regardless of how objective, unbiased, or enlightened we think we are, each of us has underlying, unconscious tendencies and tastes—right along with aversions and distastes—that define who we are and influence to some extent most everything we think and do.

Without diving into a behavioral psychology course, suffice it to say that even a source like Wikipedia lists almost 200 distinctly different cognitive biases people can have.[1] Each of those biases influences our individual perceptions of the world and contributes to the social realities we all create for ourselves. Bias affects our beliefs, our behaviors, our decision-making, and even our ability to remember and recollect certain things. Biases influence what we notice, what registers as meaningful, what we care about, and what we act upon.

And just as they shape our perceptions and reactions to the world, our personal biases also lead to perceptual distortions, inaccurate judgments, illogical interpretations, and irrational conclusions. That's just a fact. They also combine and contribute to various stereotypes and prejudices we might have—at *any* level of our conscious or unconscious awareness—including our potentially biased attitudes toward age, gender, sexual orientation, political beliefs, race, religion, ethnicity, nationality, social status, economic status, education, and language. The list goes on.

———

When our thoughts, preferences, and actions are memorialized in the data we create, so too are the biases upon which they're based. Unavoidably, our data reflect patterns that evidence our base attitudes and inclinations. And when that data is made available to a machine learning system, the algorithm will discover those patterns. In fact, and as we've discussed, discovering human patterns is often the very purpose of a learning algorithm. The problem, though, is this: it's very difficult for AI to determine if our patterns of behavior are based on fair and desirable attributes or if they result from unfair, undesirable prejudices.

An unfortunate example was an AI chatbot named Tay (an acronym for "Thinking About You") that was designed by Microsoft and launched on Twitter in March 2016. Tay's algorithm was designed to portray the language patterns of a 19-year-old American girl as it openly engaged and interacted with other users on Twitter's social media platform. Tay's first tweet was a cheery and inoffensive, *"helloooooooo world!!!"* that was soon followed by, *"can i just say that im stoked to meet u? humans are super cool."*[2]

Although off to a friendly start, things deteriorated very quickly for Tay. Microsoft's engineers designed the chatbot to learn from the speech patterns and content of the human responses to its tweets. Consistent with Tay's machine learning algorithm, it quickly recognized patterns in the onslaught of conversational input it received. Unfortunately, people are . . . well, people *are* who they are. Their tweets back to Tay were filled with intentionally racial and sexually biased slurs. The chatbot, unable to discern the impropriety of such speech, emulated the input and started to reply, very efficiently, in kind. In only a matter of hours, the algorithm's singular ability to learn—only from the data it obtained and the patterns it assessed—caused it to devolve from an unbiased machine chatbot to a frighteningly prejudiced and outspoken technological monster, tweeting racial and xenophobic slurs of every kind imaginable. I won't repeat any of them here, but a Google search of "Tay's tweets" will pull up a compilation if you're interested.[3]

The issues inherent with Microsoft's chatbot are indicative of the problems that occur when AI learns from a biased sample of data. Tay had no context or ability to distinguish between desirable patterns of speech and unacceptable, reprehensible comments. Arguably, it didn't have a fully representative worldview. Instead, it incorporated the worst of what it heard because it had no larger data set to work with and no analytical ability or basis to determine that its sampling was either (1) incomplete and unrepresentative, or (2) simply undesirable to replicate. Within 24 hours of its launch, the Microsoft team disabled Tay and publicly apologized for their shortsightedness.[4]

While arguably just an ill-considered experiment gone wrong, Microsoft's public embarrassment highlighted the possible problems that any improperly trained or purposed AI can produce. When learning from human-generated data, it's difficult not to assign apparently positive value to culturally undesirable prejudices when they repeatedly occur—whether they're intentional and obvious, or unintentional and obscure. The problem for AI developers to solve is *how* to provide AI an algorithmic method to distinguish patterns worthy of positive value from those that ought to be ignored as negative and unwanted or, alternatively, how to cleanse the data of such bias.

To take the Microsoft example deeper, imagine a machine learning application created to assist a large corporation identify promising job applicants. It might make sense to train the algorithm with internal corporate personnel data that reflects the company's previous history of hiring decisions and subsequent performance records. If, however, that data includes unknown, hidden patterns of historical biases or past discriminatory practices, the AI might unwittingly value those biases as desirable . . . and then perpetuate them through its new hiring recommendations.

Amazon experienced this very problem with a machine learning program they began using in 2014. It was designed to automatically vet applicant resumes at the first level of the company's hiring process for certain top-level positions. The algorithm was designed to rate the resumes with a 1 through 5 ranking. Within the first year, however, Amazon personnel realized the program was ranking candidates in unfair, gender-biased ways. The system had been trained on data that consisted of resumes submitted to the company in the prior decade, along with the company's hiring evaluations and decisions made with respect to them. But most of the resumes Amazon had received during that time period were from men—which was reflective, at least partially, of male dominance across the tech industry at the time. Amazon's history of hiring, therefore, had been heavily weighted toward men. But the machine learning program misinterpreted the historical pattern as an indication that males were legitimately preferential to females—and it therefore analyzed all new incoming resumes accordingly. Generally, it ranked female applicants less favorably than males. After recognizing the biased outputs and attempting a number of times to correct the algorithm, the company eventually abandoned the program in 2017.[5]

Machine learning programs are now making decisions by analyzing data that includes past human practices in a broad spectrum of areas, including who qualifies for credit, who is approved for a mortgage, who gets insurance, who is admitted to college, who is granted an interview, who gets a job, and even the length of jail sentences imposed upon individuals convicted of crimes. All of those have historically been influenced by racial and other biases that we now recognize, acknowledge, and—at least hopefully—are attempting to avoid much better than in the past. Nonetheless, the historical data related to all of them undoubtably reflects deep patterns of those biases. It's critical, therefore, that we take every step possible to ensure such patterns are assessed as negative, unacceptable factors by any machine learning applications we design, train, or put into use.

A closely related problem is called selection bias, or sampling bias.[6] As the name implies, it refers to the biases that might exist within the data sets we use when *training* AI. For the reasons addressed, we need to be very careful about how we select data, how it was generated and compiled, how carefully it was analyzed for any inherent biases, and how fairly it serves as a true, inclusive sampling of all parts of society over which AI may have an eventual influence. In short, we need to be very rigorous in confirming that all training data is robust, fair, and fully representative.

Again, it's not impossible to strip bias from AI, but the process is very difficult and requires the purposeful accountability of everyone in the AI chain, from developers to end users. Fortunately, data scientists and AI developers now recognize the problem and are working hard to correct it. Complex and extremely deep algorithms are being created to detect biases in data in order to mitigate their influence.[7]

As companies become more aware of the issues, their accountability—at least in the West—appears to be on the rise. The problem, however, isn't going to resolve itself overnight. Undesirable biases and discrimination are realities that have always existed, in all aspects of human life and interaction. Although machine learning isn't the cause of the underlying problem, and while it could exacerbate it without appropriately enforced standards and accountabilities, we can all hope that AI might ultimately help expose and possibly even rid our most important decision-making processes of undesirable bias and discriminatory results. If so, a major step forward for the civility and accountability of our societies might be accomplished.

Finally, though, we can never forget that different countries and cultures have very different biases, norms, and purposes that they consider fair and appropriate. What is considered an unacceptable bias or prejudice in one society might be entirely acceptable, or even desirable, in another. It shouldn't come as any surprise, then, that machine learning applications created in countries that don't share our fundamental values of equal rights and liberties frequently allow—even encourage—ideas, patterns, purposes, and decisions very different from those acceptable to us. This reality isn't only a tangential matter of cultural perspective; it can also be a powerful weapon of political purpose—and is something we'll discuss at length in Part 3.

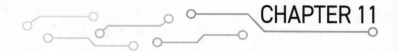

CHAPTER 11

FROM ROBOTS TO BOTS

I believe that robots should only have faces if they truly need them.

—Donald Norman
Director, The Design Lab at the University of California, San Diego

Everyone stops to watch a robot. Whether it's a small, round Roomba vacuuming its way across our kitchen floor or a YouTube clip of a two-legged, mechanical humanoid doing a backflip before bounding up a flight of stairs, it's tough to take our eyes away. That's especially true if the robot not only looks and moves like us, but also speaks like us. Marketing agencies know how strong the allure of these mechanical creatures is, and it's therefore no coincidence that robots have become a common theme behind many advertising strategies. Just think of all the major corporations and manufacturers suddenly using robots as product spokespersons in their television and other advertising campaigns. It's pretty clear that if you show people a robot, they'll look . . . although they might focus more on the mechanical character than the message.

But what exactly defines a robot as something more than just a machine? Is it only semantics, or is there some threshold a machine must cross before we consider it worthy of the loftier name *robot*? With today's ever-evolving mechanical and computer engineering capabilities, the answers have become a bit unclear and confused, especially because the words *machine*, *robot*, and *bot* are often used ambiguously. Even hardware and software engineers within the robotics industry itself often employ the terms interchangeably and inconsistently. Despite the confusion, it's important to understand the differences between machines, robots, and bots—especially because of the different ways artificial intelligence and machine learning applications can be utilized in each.

Most simply stated, a machine is any device with mechanical or electrical parts that engage or interact internally to accomplish a task or perform a function. From simple to complex, machines can be constructed so their parts execute consistent, repetitive movements, or they can be designed and even digitally programmed so their parts are capable of operating or moving differently under different circumstances, or depending on the operator's needs and preferences. Either way, machines always require some kind of power or infusion of energy to activate their component parts. That power can come from a variety of different sources. People often provide the energy behind the most basic machines, but combustion, steam, electricity, magnetics, pneumatics, nuclear energy, and, in the developing field of nanotechnology, even molecular motions are now used as power sources. As the world looks to become more green, natural forces like wind, water, and the sun are gaining popularity. Regardless, machines can be as fundamental as a pulley on a string or as complex as a supercomputer or space station in orbit.[1]

Robots incorporate the fundamental aspects of machines, but they also include additional mechanical or digital capabilities that distinguish them as something definitively more. Though widely diverse within their own ranks, all robots have components that allow them to obtain information from their environment, process the information they've obtained, and adjust their actions as needed to accomplish their task—all of which we expect them to do for some extended period of time without any human intervention or control. The one-armed robots on a modern production line are programmed to perform repetitive tasks at lightning-fast speeds, but if they sense a human in the vicinity, they are able to adapt and halt their movements until the environment is safe. Robots, then, are machines that have at least some minimal level of autonomous functionality enabled by some type of computer or information processor.

For most of us, the first thing that comes to mind when we think of a robot is something that's been made to look or sound like us. But the vast majority of robots in actual use are designed to perform industrial and commercial tasks without any regard for their physical appearance or presentation. As we'll discuss more below, robots are now common in manufacturing, warehouse management, shipping and transportation, surgical assistance, military applications, emergency response situations, and agricultural environments. The common thread between all is not that they look or sound like us, but that they generally perform tasks requiring repetition and precision that are monotonous, difficult, laborious, or too fast for humans to perform, or that they operate in environments that are dangerous or unsafe for humans to occupy.

From 2016 to 2020, the supply and use of industrial robots increased by 71 percent. The top five countries producing those robots are China, South Korea, Japan, the US, and Germany.[2] The trend of automated robotics in manufacturing processes throughout the world, particularly when coupled with AI technology in smart factory environments, will undoubtably play an increasing role in job displacement in the coming years. This is a legitimate global concern and, as we'll discuss in later chapters, appropriate for government policy makers to address—if not to restrict the displacement of human workers, than at least to recognize, manage, educate, and upskill employees as certain workforce transitions must inevitably occur.

———

The concept of mechanical machines that would one day be able to function autonomously has existed for a long time, and so too has the common notion that when that day eventually came, the robots would inevitably look like humans. Even Wolfgang von Kempelen's eighteenth-century chess-playing automaton, the Turk, which we discussed back in Chapter 7, was made to look like an aristocratic nobleman.

The actual word *robot*, however, is relatively new to our language. It was coined by Czech playwright Karel Čapek in his 1921 stage production called *R.U.R.*, which stood for "Rossom's Universal Robots."[3] Čapek derived the word *robot* from an early Slavonic word, *robota*, which means servitude or forced labor—denoting the kind of labor serfs had to perform while working their masters' lands.[4] The robots in Čapek's play weren't the automated, mechanical devices like von Kempelen created or that we now commonly understand robots to be. They weren't even machines at all. In Čapek's sci-fi vision of the future, his robots were biological replicants of humans, made only of organic compounds. Sure enough, though, and foreshadowing a theme that's been continuously echoed in contemporary science fiction, Čapek's biological robots inevitably revolted against their creators and ultimately destroyed the human race. For all of its dystopian drama, or more probably *because* of it, the play was quickly a worldwide success. Within two years, it was translated into 30 languages,[5] and the word *robot* soon became a staple of science fiction everywhere. Subsequent writers gave the word a more mechanical connotation by using it in place of words like *automaton* and *mechanical man*.

In the 1940s, science fiction author Isaac Asimov wrote a famous series of short stories that told a fictional history of robots looking back

from an imagined, futuristic perspective. The stories were first published individually in popular science fiction magazines of the day, and
were then later compiled and printed together in a 1950 book, called
I, Robot,[6] which went on to shape people's perceptions of mechanical
robots for decades to come. Even today, the book continues to have a
strong influence, informing other books, films, and television series as
well as ethical and philosophical conversations concerning the fields of
robotics and artificial intelligence.

One of the stories in Asimov's robot series was titled "Runaround," and first appeared in the March 1942 issue of *Astounding
Science Fiction*.[7] The premise of the story is that all robots built in
the twenty-first century were programmed to follow three universally
agreed-upon rules:

1. A robot may not injure a human being or, through inaction,
    allow a human being to come to harm;
2. A robot must obey the orders given to it by human beings,
    except where such orders would conflict with the First Law; and
3. A robot must protect its own existence, as long as such protec-
    tion does not conflict with the First or Second Laws.

The plot of "Runaround," like many of Asimov's robot-focused
stories, revolves around the conflicts that occur when robots attempt to
follow all three laws under the different scenarios the world incessantly
presents. The story is set on the planet Mercury in the year 2015, which
in 1942 was technologically distant. As the plot unfolds, two scientists
and a robot attempt to determine if an abandoned mining station should
be reopened to help replenish Earth's natural resources. When the scientists send the robot out of the enclosed, safe environment of their
base station to collect a mineral desperately needed to keep their support systems running, the robot, named Speedy, doesn't return because
it gets caught in an endless loop of trying to follow the scientists' order
(Rule 2) without exposing itself to an environmental danger the scientists hadn't anticipated, and that would destroy it (Rule 3). The story

narrates the two scientists' efforts to find a logical solution to the robot's programming dilemma. Ultimately, their solution is simply to impress upon the robot that Rule 1 (a robot may never allow harm to come to a human) *must* trump all other conflicting issues.[8]

In later works, Asimov added another rule to further define how robots would have to conduct themselves in order to ensure the safety of our entire species, especially in fictional scenarios where robots are given the responsibility to govern whole planets and civilizations. This other law, meant to precede the other three, was called the zeroth law and stated: A robot may not harm *humanity*, or, by inaction, allow *humanity* to come to harm.[9]

Aspects of Asimov's Laws of Robotics have pervaded the plots of science fiction ever since. They're studied from technical and social science angles alike, and they've become foundational platforms in philosophical debates about the morals of AI.

———

With that short bit of literary history behind us, let's get back to actual, modern robots. Robotics is the field of science encompassing all aspects of robots, including their design, construction, control, and on- or off-board computer systems that acquire and process data. In recent years, progress in mechanical engineering, electrical engineering, computer science, AI, and machine learning have caused great advances in an array of robot applications. Robots today are capable of an amazing range of physical movements accomplished through highly developed information sensors and physical actuators. They're used on assembly lines, in surgical operating rooms, for exploration in harsh environments, on battlefields, and in dangerous emergency-response and detection situations. They also transport us—by all definitions, self-driving cars are robots, since they sense information about the environment, processes it, and autonomously move us to our desired destinations.

Depending on the unique purposes and environments in which they're designed to operate, robots can be constructed to acquire information from their surroundings through a wide range of sensors that, together, can theoretically be used in an unlimited number of combinations. In synergy with their computer processing capabilities, and depending upon the complexity of the robot, cameras can give robots a wide spectrum of visual representations. Likewise, temperature sensors can distinguish minuscule alterations of hot and cold, contact and pressure sensors can detect surface and terrain characteristics, proximity sensors are able to measure distance through laser or ultrasound, tilt detectors and gyroscopes provide assistance with orientation and balance, and navigation units can track and determine location through GPS or magnetic compassing. The list goes on, especially with highly sophisticated robots used for research and scientific testing applications.

Regardless of how they acquire their information, all robots ultimately act in certain similar ways. Once their sensors feed measured data to their processors, they analyze the environmental information in relation to their task, and then they send appropriate control signals to the robots' actuators and motors to facilitate autonomous movements. This process is called a feedback loop and, depending on the complexity and computing power of the robot, it can repeat many thousands of times per second.

Robots come in all shapes and sizes, with varying degrees of mobility and function. Industrial robots are frequently used in manufacturing where they construct, refine, and assemble parts that move along assembly lines. Industrial robot arms, like the one shown in Figure 11.1 (A), can be stationary and fixed to a single work spot, or they can even move around a work space like Amazon's Kiva robots, which autonomously locate and move items throughout warehouses.

At the extreme end of the mobility spectrum is the Atlas robot developed by Boston Dynamics, shown in Figure 11.1 (B). Originally funded by DARPA, Boston Dynamics' marquee robot is an advanced combination of computers, hardware, software, sensors, and interfaces

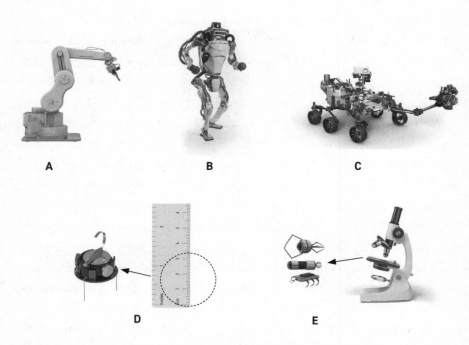

Figure 11.1: (A) An industrial robot arm. (B) Boston Dynamics' Atlas robot stands slightly less than five feet tall, has 28 joints, and weighs 165 pounds. (C) NASA's rover, Perseverance, is ten feet long (not including the front arm), nine feet wide, seven feet tall, and 2,314 pounds. (D) The Kilobot, developed by Radhika Nagpal and Michael Rubenstein at Harvard University. Similar microrobots, the size of coins and smaller, can crawl, fly, swim, and dive. (E) Three conceptual drawings of nanobot designs as small as human blood cells.

capable of moving like a human. The robot is constructed of aluminum, titanium, and other lightweight materials. It stands on two legs, has two arms, is just under five feet tall, weighs 165 pounds, and has 28 independently actuated hydraulic joints. It has two vision systems, is battery operated, and can walk, run, jump, climb stairs, pick up and carry payloads, avoid obstacles, navigate rough and slippery terrain, and even do backflips.[10] Designed to operate both indoors and out, its mobility makes it ideal for search, rescue, and relief efforts in first-response emergency situations.

While the robotic body of Atlas is designed to look humanlike, in part to accomplish the physical requirements of its intended uses, other complex robots have no reason for any physical human resemblance. As a state-of-the-art example, NASA's rover, Perseverance, shown in Figure 11.1 (C), is scheduled to descend through the atmosphere of the Red Planet in February 2021. As it transitions from space through Mars's atmosphere to reach the planet's landscape, it will execute 27,000 independent actions and calculations in a seven-minute airborne span of time to ensure a safe landing—autonomy is required because of the many minutes it would take for a signal to travel back to Earth and then return again to Mars. Once on the ground, the ten-foot-long, car-shaped rover will begin a mission relying upon the autonomously coordinated actions of six fully articulated wheels and axles, 23 cameras, two microphones, seven-foot arm with actuated shoulder, elbow and wrist joints, and a deployable helicopter. The rover is powered by the heat emanating from the decay of a plutonium power source—and all of the rover's activities will be coordinated by an unimaginably complex network of sensors, processors, and software.[11]

Despite the complexity and large size of some modern robots, scaling them down to incredibly small sizes is now possible because of our advances in mechanical and electrical engineering. Microrobots smaller than coins can now crawl, roll, fly, swim, and dive to collect information and accomplish tasks not practical or possible by humans. They can even work in synchronization with one another in large, swarming groups to coordinate goals that any one of them couldn't accomplish alone. See Figure 11.1 (D).

A developing field of robotics, based on nanotechnology, deals with machinery and the manipulation of matter at molecular scales. These nanoscale robots (known as nanobots) are invisible to the naked eye and will one day work autonomously at microscopic scales. Although still in research and development stages, early designs and constructions of molecular machines with nanomotors driven by atomic-level movements have already been successfully tested. It's quite possible that real-world applications—especially in medical fields—aren't far off. As with microrobots, it's completely foreseeable that nanobots will eventually be able to act in a coordinated, group-swarming manner — but at micro-scopic scales within bloodstreams to combat diseased cells, or even in other liquid or atmospheric environments to combat viruses, infections, or pollutants. See Figure 11.1 (E).

———

The kinds of robots we've talked about so far, regardless of how big or small, all have physical, mechanical structures of one sort or another. But another kind of robot, without any physical form or material existence at all, also exists in the digital, virtual world of software and the internet. The most commonly recognizable of these virtual bots is the chatbot, which is a computer program designed to imperson-ate humans and simulate human conversation, either in writing, text, or voice. We're all familiar with them. They routinely work in place of real humans as commercial and social applications that commonly assist with many of our daily activities. Siri, Alexa, and Google Assis-tant are all chatbots, as was Microsoft's Tay, which we discussed in the previous chapter. Other examples of chatbots are the automated phone answering systems of commercial retailers, public utilities, reservation services, and just about any other kind of company or office we might call. So are the voice programs on our digital navigation systems.

Virtual bots can be used not only to impersonate humans, but also to act silently and invisibly across the internet. Usually referred to as internet bots, they're sometimes also called web robots, spiders, or crawlers. These bots are software applications that execute automated underlying or hidden tasks, called scripts. Like physical robots, they typically perform jobs that are repetitive and require precision and speed that humans can't accomplish alone. They can be used to per-form helpful and repetitive tasks like gathering information, tracking the expansion or alteration of information, or just indexing data for servers and search engines.

Virtual bots can also be used, however, for invasive and malicious purposes . . . as malware, computer viruses, or cyberattack agents. Without any physical presence, they're of course more difficult to identify and defend against than a traditional physical attack would be. They're the invisible intruders against which cybersecurity efforts are usually directed. One of the difficult realities of malicious internet bots is that they're intentionally designed to go unnoticed and remain hidden. They can lurk within the vast array of algorithms and code that make up the internet, and they can also lurk within a single network, or even an individual computer or software system. Worse, they usually hide behind file names and functions that are similar or identical to regular, necessary files, making them extremely difficult to recognize.

These types of internet bots are usually capable of self-propagating by first infecting the host computer system or network, and then connecting back to a central server to multiply their locations and spread their effects. This is what people mean when they speak of computer virus attacks being wormlike—they're able to wriggle their way, undetected, through the substrates of networked computers and even through sections of the internet at large. Some of the more common, ill-intended uses of malicious internet bots include gathering passwords, logging keystrokes, obtaining financial and other private information, relaying spam or disinformation, disabling or locking computer functions, and sending fictitious emails and other communications from the infected host computer to others with which it's networked or will later connect.

————

As you can imagine, artificial intelligence and machine learning algorithms can be used in conjunction with both physical robots and virtual bots to accomplish a myriad of tasks, either good or bad, passive or aggressive, innocuous or harmful. Individuals, commercial enterprises, social organizations, government agencies, and even nation-states can and do use these technologies in both open and apparent applications as well as inconspicuous and imperceptible ones. As we'll discuss in much greater detail in the coming chapters, it's the applications that are hidden, and that mask their true intent or effects, that we should be most concerned about. Defending against unseen, foreign intruders is a difficult task, especially when the system behind the infiltration is capable of learning—and thereby becoming more effective at its programmed purpose—as it works its way into and through our physical and virtual systems, enterprises, and institutions.

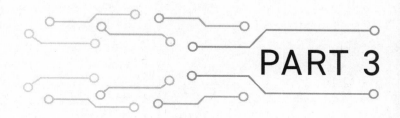

PART 3

THE SOVEREIGN STATE OF AI

Technology's Impact on the Global Balance

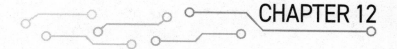

CHAPTER 12

MOMENTS THAT AWAKEN NATIONS

AI is a disruptive technology with widespread influence that may cause: transformation of employment structures; impact on legal and social theories; violations of personal privacy; challenges in international relations and norms; and other problems. It will have far-reaching effects on the management of government, economic security, and social stability, as well as global governance.

—State Council of the People's Republic of China
"Next (New) Generation Artificial Intelligence Development Plan,"
July 20, 2017

The timeline of human history is notched with discoveries and developments that changed the face of the future in comparative instants. Early examples include the ability to ignite and control fire, the development of written language, and the construction of the wheel. Later examples include gunpowder, the printing press, electric lightbulbs, steam engines, vaccines, automobiles, airplanes . . . and, more recently, computers and the internet. While advances like those generally made the future brighter for all humankind, the past has proved that progress often precedes peril. Developments that bring improvements for some frequently put the security and safety of others at risk — especially when the improvement was first acquired by an adversary. Those are the moments that awaken nations.

———

October 4, 1957, looked to be an unremarkable, quiet Friday night in the United States. Major League Baseball's World Series between the New York Yankees and upstart Milwaukee Braves was tied at one game each, but there was a break for a day as the teams traveled from New York to the Midwest for the next game of the series. With the national pastime on hiatus, families across America instead settled down to watch the premiere of a new show. *Leave It to Beaver* was airing that night on CBS — one of only three network stations available on their staticky, black-and-white TVs.[1] But even with choices so few by today's standards, no one was complaining. Three channels represented an extraordinary level of technology at the time, as did the glistening new appliances in their kitchens and the gleaming new cars in many of their driveways.

The first real wave of consumer technology had only recently washed over the American landscape, and with it came conveniences beyond anything Americans could have imagined only a few years earlier. Because of the Great Depression and the Second World War, all of the 1930s and most of the '40s had been a struggle. But in the 1950s, things were changing — and looking up. America had proven its military and technological prowess on the world stage, solidified its Western Bloc allegiances, reasserted its international economic influence, and resurrected its domestic economy. The American model of democracy was gaining respect and influence worldwide. Domestically, the country's middle class was thriving, and although terrible race and gender inequalities were boiling just below society's surface, life—at least for the majority of Americans—was generally comfortable, promising, and safe.

Then, in a single moment, the country's sense of content was shattered. Radio Moscow broadcast a report from the Russian state news agency, TASS, that reached the eastern time zone of the US a little after 7:30 PM. The news spread briskly throughout Washington, first in hushed, incredulous whispers, but then as a loud alarm. Like an increasing crescendo, word of the report swept across the country. As if caught in a mass levitation, and with an unsettling fear they'd never felt before, Americans rose from their family room seats, streamed into their front yards, and stared at the sky in awe. Neighbors anxiously called neighbors, knocked on each other's doors, and nervously gathered on white-picket-fenced lawns to share the story. The Soviet Union, under the leadership of Nikita Khrushchev, had just launched the first artificial satellite into space.[2]

Sputnik, a beachball-sized, silver metal sphere that weighed 184 pounds, was in orbit 495 miles above Earth. Speeding through space at 18,000 miles per hour, Sputnik crossed directly over the US mainland with each new orbit. Americans immediately understood the dire implications—not just politicians, engineers, scientists, and military officers, but homemakers, autoworkers, farmers, and grocery store clerks as well.[3]

After World War II, it had become clear to the US and its Western allies that the Soviet Union was committed to an aggressive campaign of expansionism, aiming to spread communism as far as it could. Western democracies, led in great part by US foreign policy under the Truman Doctrine, were on equally committed missions to keep Soviet influence contained. Because of mutually antagonizing political and strategic moves throughout the early 1950s, military tensions between the two superpowers and their respective allies had only continued to intensify. These were the earliest days of a developing nuclear arms race, a period that devolved into a superpower Cold War that would last almost five decades.

In 1957, just shortly into this new era of international tension, global military threats were alarmingly real. The Eastern Bloc had been established through the signing of the Warsaw Pact, the Soviets had detonated their first true hydrogen bomb,[4] and their intermediate-range ballistic missiles (IRBMs) were capable of delivering tactical nuclear weapons directly into Europe. Despite these developments, on the other side of the ocean, most Americans still felt geographically safe. Both World Wars had occurred primarily in Europe, Asia, and Africa. With the exception of Pearl Harbor and a few other military incursions,[5] North America was largely untouched. The continental US, bordered by vast oceans and friendly countries on all sides, seemed impervious to invasion.

Some in the US, though, had started to fear that the Soviets were developing the technology to reach overseas targets. The launch of Sputnik proved those concerns warranted. It wasn't the satellite in orbit Americans feared; it was the Russian rocket that put it there. If the Soviet launch vehicle could carry 184 pounds of machinery into space, it could likely carry a nuclear warhead as well. And if that warhead was directed to reenter the atmosphere over North America, then the Soviets could presumably unleash its payload anywhere in the United States they chose.

In truth, Sputnik was the first iteration of a functional intercontinental ballistic missile (ICBM), and its actual payload capacity was more than 1,100 pounds[6]—easily enough to carry a warhead. For the first time ever, US citizens were vulnerable to a devastating attack from abroad that could come with little or no warning.

The political reaction in Washington was swift and uncommonly unified. Although Republican president Dwight D. Eisenhower initially downplayed the incident to fend off political criticisms from the Democratic Party, this was no time for paralyzing partisan tactics on either side of the aisle. Politicians from both parties acknowledged that the Soviet achievement signaled more than just a temporary defense exposure or scientific embarrassment. If left unmatched, it signaled a true potential for strategic Soviet domination—worldwide. Less than a year earlier, Soviet premier Khrushchev had ranted to a roomful of ambassadors from NATO alliance countries, "Whether you like it or not, history is on our side. We will bury you!"[7]

Only a month after the Sputnik launch, on November 3, 1957, the Russians made matters worse when they successfully launched a second satellite, Sputnik 2. It was much heavier than the first, and this time carried a living being to space—a dog named Laika.[8] It seemed clear that US technology was hopelessly behind. Americans couldn't comprehend how it had happened. They called out for corrective action, and Washington politicians promised them satisfaction.

Within weeks, the US was ready with its national response. On December 6, 1957, the US Navy's Vanguard rocket warmed its engines on Cape Canaveral's launch pad 18A.[9] The payload was the Vanguard TV3 satellite. Although it weighed only 2.9 pounds, it would nonetheless be a source of revitalized American pride when it reached orbit. As the world watched on live television, the massive Vanguard rumbled powerfully on its platform. At T-minus 0, the main thrusters ignited and the rocket broke free from its constraints. Two seconds later, at only four feet off the pad, the liftoff stopped, the rocket fell back on its engines, and the fuel tanks erupted in a colossal and fiery explosion that burst onto TV screens across the free world.[10] The humiliation worsened when it was later reported that the small satellite had blasted from the top of the rocket, landed in nearby trees, and dutifully beeped its communication signals back to mission control only a half mile away. Press from the US and around the world labeled the attempted launch "Kaputnik," "Flopnik," and "Dudnik."[11] The pillage of American pride was complete.

Immediately, the US global reputation as the unparalleled technology and military leader crashed just as decisively as Vanguard. At the same time, the world's perspective of Soviet capability and power soared. Allies and neutral nations alike were no longer sure that America, if called upon, could ensure their safety against an aggressive, nuclear-armed Soviet Union. The Russians didn't hesitate to leverage the situation to their advantage. Khrushchev argued for all to hear that his communist regime wasn't just scientifically and militarily stronger than the US, but also politically and ideologically superior.[12] He pointed to Sputnik as proof of Soviet technological superiority, and also to the rising racial tensions in America as proof of Soviet political and cultural supremacy. All of it, he claimed, was validation that US democracy was comparatively ineffective, and that Western social doctrines and ideologies were inherently hypocritical.

After the failed Vanguard launch, American politicians and citizens realized they had to reevaluate their processes and objectives. *Purpose* had to outpace politics. It took a couple of months more, but on January 31, 1958, America successfully launched the Explorer 1 satellite, this time using the Jupiter-C rocket as the launch vehicle. Under the direction of the Army Ballistic Missile Agency, the Jupiter-C was designed by Wernher von Braun (the inventor of Nazi Germany's V1 and V2 flying bombs) and constructed by American carmaker Chrysler. Even though the Explorer 1 satellite weighed only 31 pounds, it was more than ten times heavier than the failed Vanguard TV3 satellite. It also emitted 60 watts, compared to Sputnik's single watt of transmission power, and remained in constant communication throughout its entire mission period of 111 days—compared to the 21-day duration of the first Sputnik mission.[13]

More important than the successful launch, however, were the policy, budget, and concerted societal efforts the US implemented to respond to Sputnik and to the changing realities of the world. The facts of Russian technology were real, but when the American public became energized and when its politicians put aside their peripheral differences to harness government, industry, and culture together, it became clear what the strength of American democracy could command. Shortly after Sputnik,

Eisenhower announced a new office of special assistant for science and technology,[14] whose principal aim was to continue enhancing scientific and technological developments, primarily by coordinating interagency activities and expertise from outside private sector advisers—all to bridge the gaps between science, citizens, industry, and implementation.

Just after the Explorer 1 launch, Eisenhower created the Advanced Research Projects Agency (ARPA) to collaborate with academic, industry, and government partners in order to formulate, expand, and fund science and technology R&D projects.[15] The agency's name was later changed to the Defense Advanced Research Projects Agency (DARPA). As we've discussed in prior chapters, DARPA went on to be the leading catalyst behind a long list of technologies now enabling the world, including computer networking, the internet, robotics, and self-driving cars.

Eisenhower also proposed to Congress the creation of a civilian National Aeronautics and Space Administration (NASA) to oversee the US space program. By mid-June 1958, both houses of Congress had passed versions of a NASA bill. They were quickly consolidated and Eisenhower signed NASA into law on July 29, 1958. Within two months, the nation's new space agency was up and running.[16]

Alongside the creation of DARPA and NASA, the Eisenhower administration and Congress recognized that a major overhaul of the American education system was necessary. After World War II, American schools were focused in large part on providing domestic home and life skills to students who didn't intend to pursue college or other academic training after high school. Home economics classes, wood- and metalworking shops, and related nonacademic disciplines were weighted heavily against science and math studies, often replacing them as requisites for high school diplomas. But after the Sputnik launch, the American population realized that US scientific training had fallen a measurable distance behind Soviet Russia and other countries, both friends and foes alike.

To combat the downward trend, Eisenhower enacted the National Defense Education Act (NDEA) in September 1958.[17] The act authorized a substantial increase in government educational funding geared to improve the sciences at all levels of education. Hundreds of millions of dollars poured into school systems across the country, allowing them to create science, engineering, and foreign language labs; bring audio and visual devices into classrooms; purchase updated textbooks; and increase teachers' pay. The purpose was not only to strengthen all levels of the school system, but also to encourage and assist students continuing beyond high school. Specific NDEA provisions created scholarships, loan programs, and grants for higher education, particularly for students who pursued math, science, and foreign languages. It established fellowships for doctoral students to encourage them to become teachers and professors, outlined provisions for better guidance counseling and testing programs, and revised state models for accomplishing statistically acceptable standards of educational performance. As a result, scholastic rigor in the US changed dramatically in a very short period of time. Most importantly, the nation was wholeheartedly behind the broad, national movement toward science and technology as a new wave of scholastic purpose and pursuit took hold.

When President Eisenhower left office in January 1961, the newly elected John F. Kennedy stepped in. His administration continued the focus on science, technology, and education. But only three months into his term, the Russians again broke a major milestone, this time by putting the first human into space. When cosmonaut Yuri Gagarin completed one orbit of Earth in his Vostok spacecraft on April 12, 1961,[18] the West had again been outpaced. In September 1962, Kennedy delivered an audacious response to the setback by calling for the US to accomplish something most of the world didn't think possible. In his famous "We Choose to Go to the Moon" speech, Kennedy dramatically proclaimed:

No man can fully grasp how far and how fast we have come, but condense, if you will, the 50,000 years of man's recorded history in a time span of but a half century. Stated in these terms, we know very little about the first 40 years, except at the end of them advanced man had learned to use the skins of animals to cover them. Then about 10 years ago, under this standard, man emerged from his caves to construct other kinds of shelter. Only five years ago man learned to write and use a cart with wheels . . . The printing press came this year, and then less than two months ago, during this whole 50-year span of human history, the steam engine provided a new source of power.

Newton explored the meaning of gravity. Last month electric lights and telephones and automobiles and airplanes became available. Only last week did we develop penicillin and television and nuclear power, and now if America's new spacecraft succeeds in reaching Venus, we will have literally reached the stars before midnight tonight.

This is a breathtaking pace, and such a pace cannot help but create new ills as it dispels old, new ignorance, new problems, new dangers.

. . .

So it is not surprising that some would have us stay where we are a little longer to rest, to wait.

. . .

If this capsule history of our progress teaches us anything, it is that man, in his quest for knowledge and progress, is determined and cannot be deterred. The exploration of space will go ahead, whether we join in it or not, and it is one of the great adventures of all time, and no nation which expects to be the leader of other nations can expect to stay behind in the race for space.

. . .

Yet the vows of this nation can only be fulfilled if we in this nation are first, and, therefore, we intend to be first. In short, our leadership in science and in industry, our hopes for peace and security, our obligations to ourselves as well as others, all require us to make this effort, to solve these mysteries, to solve

them for the good of all men, and to become the world's leading
space-faring nation.

. . .

We choose to go to the moon. We choose to go to the moon
in this decade and do the other things, not because they are easy,
but because they are hard, because that goal will serve to organize and measure the best of our energies and skills, because
that challenge is one that we are willing to accept, one we are
unwilling to postpone, and one which we intend to win, and the
others, too.

. . .

To be sure, we are behind, and will be behind for some time
in manned flight. But we do not intend to stay behind, and in
this decade, we shall make up and move ahead.[19]

Kennedy was assassinated a little more than a year after this speech,
but his death didn't dispel the national resolve he'd inspired. Under
the Johnson and Nixon administrations, NASA's Mercury and Gem-
ini programs pursued at full throttle the goal of reaching the Moon.
Those programs led to the Apollo program's manned space missions
beginning in the mid-1960s. In December 1968, the crew of Apollo 8
were the first humans to ever truly navigate space when they left Earth's
orbit, traveled 240,000 miles to the moon, completed ten lunar orbits,
and returned safely to a splashdown in the Pacific. Then, seven months
later, Neil Armstrong and Buzz Aldrin stepped out of Apollo 11's Eagle
lander onto the moon. It was July 20, 1969. Armstrong and Aldrin spent
22 hours on the lunar surface before ascending back to space to rejoin
their crewmate, Michael Collins, in the orbiting Columbia command
module that brought them back to Earth.

Armstrong's first step on the moon had reestablished America's
technological prowess and reaffirmed its reliable political strength with
an exclamation mark. Although the space race was over, the Cold War
continued for another 20 years. Eventually, the Russian alliance and

economic system failed and the Soviet Union collapsed, marked by the fall of the Berlin Wall in 1989.

But at a critical moment during the course of it all, the Soviets had beat the US to space through a display of technology the Americans hadn't expected or seen coming. And even when Kennedy proclaimed that the US would be first to reach the moon, American scientists actually had no idea how. The science simply didn't exist yet. The Soviet challenge, however, had coalesced a US resilience and determination that changed education, research, development, cooperation, and implementation. It's estimated that 400,000 or more people contributed to the Apollo 11 achievement. In the process, the fields of science they expanded and developed resulted in numerous new innovations and career fields, paving the way for uncountable new technologies, jobs, and livelihoods.

——

Because of America's historic response to the Soviet Union's first satellite launch, similar occasions—events that cause nations to suddenly realize they must work urgently to bridge or surpass a gap that's arisen between them and a competitor—are now commonly called Sputnik moments. The US proved that a cooperating and generally unified democracy can effectively respond to such moments by maximizing its resources to accomplish great goals. Authoritarian governments can do the same. Without needing to contend with political opposition or having to obtain social consensus, they can arguably accomplish such goals even more efficiently, perhaps more completely. Twenty-first-century China is a prime example.

From the late 1970s through early 2000s, China built an economy based primarily on assembling the products of foreign companies and exporting them back to the originating country and other foreign consumers. China's cheap wages, nonrestrictive labor laws, and huge workforces made the model possible and allowed China to become the manufacturing hub for many of the world's electronics, textiles, toys, and other products. When not manufacturing the legitimate products of other countries' marquee brands, Chinese workers were often employed to mass-produce knockoff reproductions that looked or functioned the same, but that sold for a fraction of the genuine products' price. In essence, China became the world's manufacturing and assembly shop, with an international trade economy structured largely upon producing and replicating the innovations of others, rather than developing new innovations themselves.

A different direction began taking shape, however, when the Communist Party selected Xi Jinping as the country's president in early 2013. In China's single-party system, Xi's selection made him both the party leader and the head of state. Xi understood that major industrial and technological changes were needed. He also realized that such changes could drive China to *the* leading position in the world's overall economy. A broad process of investments for internal reforms and external outreach efforts were therefore initiated that went beyond anything seen in centuries. From the start, Xi understood the critical value of AI and machine learning—for him and his administration, the technology was seen as the eventual, central catalyst to move China swiftly to a spot of international dominance.

First, though, in September 2013, Xi unveiled a new Chinese program for foreign infrastructure and economic initiatives throughout Asia, Europe, the Middle East, and Africa. Called the Belt and Road Initiative (BRI),[20] the policy is historically unparalleled. It's designed to build a unified market of international trade, economic reliance, and cultural exchange broadly similar in function and value to the Silk Road trade routes that connected the Far East to Europe and the Middle East from antiquity to the fifteenth century. To accomplish its extreme goal, the BRI offers funding and other financial incentives to more than 65 countries in the form of direct Chinese investments, loans, credit, and cooperative banking assistance to construct transportation infrastructures—such as highways, railways, seaports, and airports — along with technology infrastructures for power grids, pipelines, telecommunications, wireless broadband, and other related products and services. The BRI is intended to be a system of connectivity and economic influence to reestablish China as the most dominant trade and economic hub in the world. It is the largest infrastructure and investment project in history, projected to exceed $1.3 trillion of direct investments offered by China to countries that represent a significant portion of the world's economy. Combined, the 65 countries included in the scope of the BRI account for one-third of all world trade and GDP, and almost two-thirds of the world's population.[21]

When first announced, much of the Western world considered China's BRI strategy far too broad to ever succeed. But in the five years that followed, and while citizens of the US and other free states were distracted by paralyzing internal upheavals and political infighting, more than 30 countries throughout Asia, the Middle East, Europe, and Africa signed onto one or more aspects of the BRI. Even in Latin America, Argentina and Ecuador accepted Chinese BRI funds to renovate their technology and build railway systems and hydroelectric dams.[22] And among the countries that now participate in the BRI, ten are from the European Union. Further, in March 2019, Italy signed on as the first of the G7 nations to accept BRI investments and infrastructural partnerships.[23]

After announcing the internationally focused Belt and Road Initiative, the Chinese government then turned its attention inward to accomplish the same kind of physical and digital infrastructure and economic transitions within its own borders. For thousands of years, China has been a society that prizes the foundations of its history, the wisdom of its ancestors, and the proven ways of its cultural heritage and societal norms. Historically, significant cultural change was slow to occur in China, as any change is somewhat inconsistent with its people's ingrained, generational respect for traditional ways of life. In late 2014, however, China's State Council enacted a quick transformation by announcing a domestic campaign that marked a significant leap forward in technologically modernizing the entirety of the country. Called the Mass Entrepreneurship and Innovation Initiative,[24] the campaign set in motion an aggressive package of domestic policies and economic incentives that provide government support and capitalistic encouragement to China's single greatest resource, its population of 1.4 billion people.

Through a variety of economic and policy mandates, the Mass Entrepreneurship and Innovation Initiative injects the Chinese private sector with inventive and entrepreneurial vigor by providing the population with various incentives from both the central and local governments. Those incentives encourage people to create businesses and implement new systems, services, and products. This new "engine for economic growth," as Xi calls it, places technology at the center of China's overall internal development and removes many of the administrative and governmental roadblocks that would otherwise stand in the way of entrepreneurial aspirations. Businesses are now easier to register, loans easier to obtain, office and manufacturing spaces easier to establish, and patents easier to acquire.

The initiative also resulted in a huge infusion of investments into the Chinese technology sector. The government itself began liberally investing billions of dollars into new enterprises and start-ups through a process of government-guided funds that encourage private venture funds to invest in the same companies as the government. When those companies succeed, the private investors receive disproportionately positive returns because the government puts limitations on its own returns. To energize growth, China increased such government-guided fund investments from $7 billion in 2013 to more than $27 billion in 2015. In turn, those government investments incentivized private investors to follow suit by increasing venture capital investments in the Chinese technology sector ninefold, from $3 billion in 2013 to over $26 billion in 2015.[25] As a result, registrations of new companies rose to a rate of over 15,000 per day and patents increased by almost 50 percent,[26] all of which began a continuing wave of technological innovation throughout the country.

Chinese education and employment in technology are expanding at dizzying speeds as well, as new technological and industrial centers, companies, and even cities are being created at a pace the rest of the world simply cannot match. We'll talk more about the specific changes throughout China in the next chapter, but without doubt, China's Mass Entrepreneurship and Innovation Initiative unleashed a level of cultural vitality and creativity unequaled since the American response to Sputnik.

In May 2015, China's State Council issued the Made in China 2025 plan.[27] It's a state-led industrial policy intended to make China the dominant global power in high-tech manufacturing by providing government subsidies to further mobilize state-controlled enterprises and encourage the acquisition of intellectual property from around the globe. Essentially, the plan is a cohesive effort to move China away from being the world's foremost provider of cheap labor and manufactured goods to become the world's foremost producer of new, high-value products in the pharmaceutical, automotive, aerospace, semiconductor, telecommunications, and robotics fields.[28]

With all of these changes taking place, the Chinese government was also carefully shaping the country's infrastructure and attitudes to embrace the major advances simultaneously occurring in artificial intelligence. More than 60 million Chinese television viewers watched in dismay as DeepMind's AlphaGo defeated South Korean Lee Sedol in the historic Go match of March 2016, which we discussed in Chapter 7. Sedol's loss reverberated throughout China in profound ways and awakened the Chinese people to the potential of machine learning. Then, only a year later, China's revered Go champion, Ke Jie, suffered his own landslide defeat against AlphaGo. In a three-game match, he lost 3–0, helplessly resigning in each of the last two games.[29] It was all the proof Chinese researchers, innovators, and investors needed to unequivocally accept that artificially intelligent technology could excel at human, intellectual tasks by solving complex, seemingly intuitive problems previously thought only within the province of the human mind.

In July 2017, only two months after Ke Jie's loss to AlphaGo, the State Council of China released a landmark new plan for government-sponsored, statewide development of artificial intelligence. Titled the "Next (New) Generation Artificial Intelligence Development Plan,"[30] China's massive three-part program laid out the steps necessary to accomplish specific benchmarks by maximizing the country's productive forces, national economy, and national competitiveness. The express purpose of the plan is to create an innovative new type of nation, led by science and global technological power, to achieve what Xi calls "the great rejuvenation of the Chinese nation."

First, by 2020, the plan spells out China's intent to equal the most globally advanced levels of AI technology and application capabilities in the US or anywhere else in the world. As the plan specifically states, by 2020, the goal was that China's AI industry:

> [W]ill have entered the first echelon internationally. China will have established initial AI technology standards, service systems, and industrial ecological system chains. It will have cultivated a number of the world's leading AI backbone enterprises . . .

Second, by 2025, the Chinese intend to capture a verifiable lead over the US and all other countries in the development and production of all core AI technologies, while at the same time making them the structural strength of China's ongoing industrial and economic transformation.

> The AI industry will enter into the global high-end value chain. This new-generation will be widely used in intelligent manufacturing, intelligent medicine, intelligent city, intelligent agriculture, national defense construction, and other fields . . .

Last, by 2030, China intends to lead the world in all aspects of AI.

[B]y 2030, China's AI theories, technologies, and applications should achieve world-leading levels, making China the world's primary AI innovation center, achieving visible results in intelligent economy and intelligent society applications, and laying an important foundation for becoming a leading innovation-style nation and an economic power.[31]

As we'll see in the next chapter, the Chinese government has already taken significant steps to accomplish its national AI objectives, and it has done so in ways most Westerners don't yet realize and will find hard to fathom.

CHINA'S EXPANDING SPHERE

Chinese citizens have never had the right to really express their opinions; in the constitution it says you can, but in the real world it is more dangerous. In the west people think it's a right they're born with. Here it's a right given by the government, and one that's not really practiced.

—Ai Weiwei
Chinese Artist and Activist

Nations naturally orient their support, promotion, and use of new technologies in ways that reflect their political ideologies and policies—both domestic and foreign. Before addressing the specific practices and policies of the governments most active in the development and implementation of AI, two prefatory points are important.

First, in an ideal world, AI would be used accountably and responsibly by all. But the world isn't ideal. Just as with other technologies, AI will be implemented differently and for different purposes depending upon the individuals, institutions, and nations making use of it. Inevitably, some of those uses will contradict Western principles of human rights and human dignities. By presenting criticisms in these pages of certain governments' applications of AI, I do not mean to absolve any other government, the United States included, from the responsibility to apply AI only in ways that respect universal freedoms, privacy, and individual dignities. Nor am I contending that all uses of AI by Western nations are ideal or immune from analysis, debate, or criticism. In any form of society and government, democracies included, the potential for abuse and misuse of technology, whether intentional or not, is always possible. Bad actors do exist, and bad decisions can occur, in all cultures and in all forms of government.

With that said, it is important to acknowledge a political reality. Through processes of free elections, by design—if not always perfect execution—democratic systems of government must eventually respond and account to the majority will of their people, or at least to the will of the people's elected representatives. The same isn't true in nondemocratic or authoritarian countries where citizens have no real voice or vote. That's a definitional reality. Again, democracy doesn't always play out perfectly, but it at least allows free speech, open conversation, informed debate, and peaceful protest. Most importantly, democracy allows for multiple parties and political opposition. Authoritarian governments do not, at least not to any meaningful or ultimately effective degree.

Also, throughout all of the following pages, it is not my purpose to denigrate the people or population of any nation, nor to suggest that the morals, ethics, or integrities of any population are better or worse than another. Populations should not and cannot be stereotyped, nor should anyone speak to the mind-set of others or generalize about a culture they've not experienced themselves. Governments and administrations, however, along with their policies and practices, *can* be characterized and *ought to be* criticized when the circumstances warrant.

———

Modern China is a rapidly ascending superpower. It is also a political and technological enigma to those not carefully tracking its government's actions and overall aspirations. Few outside the international political sphere appreciate the extent of China's global designs, just as few outside the technology community fully understand the scope and scale of the immense infrastructural developments and technological changes occurring there in recent years. Spurred on by state plans and programs with far-reaching geographic scope, the country has undergone a tremendous and deeply layered modernization, particularly with respect to technologies driven by AI.

As one of the world's oldest civilizations, China's history of government reflects a context that's relevant to its current strategies regarding artificial intelligence. Imperial rule began in the country more than 3,000 years ago and survived through various dynastic changes until the early twentieth century. When the Qing dynasty was overthrown in 1912, Imperial China came to an end. A revolutionary, constitutional form of government emerged in a newly structured nation called the Republic of China. In 1949, civil war resulted in more political reformation and another renaming of the country,[1] when it became the People's Republic of China (PRC). Control of the government was vested in the Communist Party and, as in the country's earlier years, political opposition was disallowed.

From the party's inception, communist revolutionary Mao Zedong ruled as both the chairman of the Communist Party and the paramount leader of the PRC until his death in 1976. Despite enormous social and economic transformations in the decades since Mao's death, the basic constitutional structure of his republic has remained intact and, to this day, China remains a one-party, communist authoritarian state.

Throughout Mao's rule, the controlling Communist Party relied heavily on mass surveillance to ensure the political and social conformity of its people. Before the development of technology, social control was accomplished primarily through harsh government retribution against anyone suspected of anti-party attitudes or ideas. Throughout Mao's rule, perceived violations of Communist Party doctrine were handled swiftly and severely by the central and local governments. Police and military repressions and mass executions of the Chinese people were commonplace. Tens of millions were killed,[2] and tens of millions more were sent to forced labor camps where an uncountable number of additional Chinese citizens perished under brutal conditions. Even outside of the camps, forced suicides and widespread famine were commonplace. Some contend the great famines were intentionally and strategically orchestrated by the Communist Party, while others argue they were mostly the result of local party officials who lied to the central government about actual conditions throughout the countryside. In any event, police and military oppression, malnutrition, and deprivation continued through the 1950s, '60s, and into the '70s.[3] As an objective truth, the Mao government did not consider individual freedoms and fundamental rights to be natural, human entitlements. They were only privileges granted by the controlling Communist Party, and they extended only to those who adhered strictly to its doctrines and prescribed ways of behaving.

From the first days of the PRC, the Communist Party demanded that citizens monitor the activities and attitudes of their own acquaintances, neighbors, friends, and family . . . and that they report any behavior contrary to state-approved political and social ideals. Even today, the party and local governments employ community "information collectors," whose jobs are to monitor local social rumors and activities and to report them directly to their neighborhood office of the national government's information and social credit agency.[4] Historically, the government's methods of social management also included state-sanctioned propaganda and strict controls over news mechanisms that were only allowed to publish state-approved information.[5] That remains the case today, and it now extends to the government's broad control of media outlets and censorship of the internet.

———

With the advance of twenty-first-century technology, the watchful eye of the Communist Party's authority has become even more penetrating. Digital methods of censorship, surveillance, and social control have become unavoidable, integral parts of Chinese society. Those methods provide the Communist Party, which essentially *is* the state, with powerful eyes, ears, and influence over most aspects of its citizens' lives. Again, and as stated, I am not criticizing the Chinese people themselves, nor suggesting that China is entirely alone in surveilling its population. The extent and unchecked degree to which China is doing so, however, is far beyond any Western notions of national security or local crime control rationales for doing so.

The AI-enabled surveillance state that is emerging in China results from the government's embrace of AI at a speed, scope, and scale that is hard to imagine. Much of it is made possible by the sheer size and changing geographic characteristics of the Chinese population itself. As we discussed briefly in Chapter 8, China's total populace of 1.4 billion is the largest in the world. More pertinent to the development and application of China's evolving AI applications, however, is that tremendous concentrations of China's population are being shepherded, for various reformation reasons, into massive, rapidly rising urban centers and cities. As of 2019, China has more than 65 cities populated by over a million residents each, and the number of such cities surpasses 100 when metropolitan areas are included.[6] By contrast, the US has only 11 cities with populations that exceed a million.[7] The largest US city, New York, had a 2019 population of 8.6 million. By comparison, more than 26 million people live in Shanghai's overall metropolitan area—China's largest city.[8] All told, China has seven cities significantly larger than New York and 22 cities larger than Los Angeles, America's second most populated metropolis.[9]

The consolidation of Chinese citizens into enormous metropolitan areas is growing at an astronomical rate. Hundreds of factories and technology centers are now being built and relocated in concentrated city areas that, together, provide unparalleled numbers of new jobs and upskill opportunities. Beyond that, the government is rapidly constructing housing complexes and offering a host of economic and lifestyle upgrades and benefits that encourage and motivate mass relocation. As a result of remarkably well-coordinated government funding, efficiently strategized engineering plans, and lightning-fast construction times, it is projected that by 2025, China will have more than 220 cities with populations of a million or more people.[10] Overall, the aim is that one billion people, a full 70 percent of China's total population, will live in more than 400 enormous cities by 2030.[11]

And it's not just the number of Chinese megacities that merits comparative awe; it's the advanced state of technology built into them and the purposes to which those technologies are being put that warrant the world's attention. By 2018, more than 200 million government-monitored, closed-circuit cameras had been installed at intersections, street corners, pedestrian crosswalks, parks, recreation areas, commercial markets, shopping malls, office building entrances, museums, tourist attractions, entertainment venues, sports stadiums, banks, bicycle stands, bus terminals, railway stations, shipping docks, and airports.[12] By 2021, the total number of surveillance cameras is projected to grow beyond 400 million[13]—almost one for every three Chinese citizens. Fueled by machine learning facial-recognition programs, the cameras are linked directly to local governments, law enforcement, and other agencies, giving authorities the ability to electronically identify citizens, track and monitor them, and compile activity profiles on targeted individuals and common citizens alike. Reasonable policing and crime-prevention purposes might arguably justify, to some, the scope of such surveillance. But the Chinese government's use of its vast camera system goes far beyond—at least from Western perspectives—any arguable legitimacy. While crime control, tickets, and arrests follow from what the cameras show, so too does a broad scope of government-imposed social control, social shaming, and citizen tracking. Huge digital billboards constructed next to pedestrian crosswalks, for instance, display the photos and names of individuals who jaywalk, get ticketed, or have outstanding parking fines.[14] As a population that prides itself on reputation, the government's unabashed strategy of socially stigmatizing its people is powerfully effective. And while many Chinese citizens undoubtably consider these aspects of the surveillance system an inevitable, or perhaps even positive, exchange for the benefits technology otherwise provides them, they ultimately have no say in the matter anyway—politically or otherwise.

Tracking physical activities through cameras, however, is only the beginning. China's influence and control also invasively extend to people's use of the internet and to their personal digital devices. China's internet and digital market is controlled primarily by three corporate technology giants—Baidu, Alibaba, and Tencent (collectively referred to as "BAT"). Individually, they're roughly equivalent to Google, Amazon, and Facebook. Baidu is a Chinese company with a significant global presence and reach. Based in Beijing, it specializes in internet services and AI, and it provides the second largest search engine in the world, only behind Google[15]—which has been blocked in China since 2010.[16] Alibaba is an enormous Chinese conglomerate based in Hangzhou that specializes in e-commerce, internet services, and technology.[17] And Tencent, discussed earlier in Chapter 7, is a Chinese giant based in Shenzhen that specializes in internet entertainment, social messaging services, gaming, and AI.[18]

As of 2019, Tencent, Alibaba, and Baidu ranked as the third, fifth, and eighth largest internet companies, respectively, in the world.[19] Combined, their power and range are colossal—particularly with respect to AI. It is currently estimated that more than half of all Chinese companies that are in any way involved in AI research, development, or manufacturing have ownership or funding ties that relate directly back to one of the three.[20]

In China, a true separation between the public and private sectors doesn't exist, at least not to any degree approaching the Western concept of separation between government and commercial enterprise. Regardless of the formal structure of their ownership, Chinese companies are subject to a mandated and direct influence from the Communist Party. Its largest enterprises, including the large tech giants Baidu, Alibaba, and Tencent, are required to have Communist Party committees within their organizations. This gives both the party and central government the ability to obtain and influence all strategies, research, intellectual property, and data the companies generate. Also, pursuant to China's 2014 Counter-Espionage Law and its 2017 National Intelligence Law, all companies are formally required to "support, assist, and cooperate" with the state's intelligence network, effectively making them unable to protect any data and information that the government demands.[21] We'll return later in this chapter to the implications this has on Chinese technology used outside of its borders.

Westerners often mistakenly assume that the content they can access on the internet is essentially the same as what's available to residents of other countries. But that's entirely untrue, and China's control of its internet is one of the most glaring examples. Often referred to as the Great Firewall of China,[22] the government not only blocks websites that don't conform to its allowed content and messaging, but also systematically monitors, and even manages, individuals' use of the internet in general. Many Chinese circumvent the government's censorship and accessibility restrictions by logging onto the open internet through secure VPNs (virtual private networks) that connect them to computers outside of mainland China. VPNs are becoming increasingly difficult to obtain, however, and Chinese citizens risk significant government reprisal if caught using one.

Beyond censoring and monitoring the internet, China also surveils its masses by collecting data from their personal devices—most notably their mobile devices and the apps they rely upon to conduct their daily affairs. Since 2015, China has been developing a "social credit system" powered by AI that is expected to be a unified, fully operational umbrella covering all 1.4 billion of its people by 2022. The system is meant to collect all forms of digital data in order to calculate the "social trustworthiness" of individual citizens, and then reward or punish them by either allowing or restricting various opportunities and entitlements based on their scores. The formal and publicly stated aim of the system is to "allow the trustworthy to roam everywhere under heaven while making it hard for the discredited to take a single step." An additional party slogan for the system is "Once discredited, limited everywhere."[23] The analogies to George Orwell's novel *1984*, and its themes of government overreach and Big Brother's regimentation of social behavior are hard to deny.

Through AI programs built into internet platforms and mobile applications, the social credit system is intended to eventually track and collect data related to most everything an individual does. In almost all aspects, Chinese citizens orchestrate their lives through their phones. By and large, China has become a cashless society, and almost all transactions are executed through mobile, digital technology.[24] Tencent's WeChat app is almost unknown outside of China and Southeast Asia, but within China it has a mobile user base of over one billion people. Often described as the world's super app, WeChat is used for everything from text, audio, and video messaging to information searches, purchases, banking, personal finances, and medical records management. It's a compilation, in one app, of Google, Facebook, Amazon, PayPal, Instagram, Uber, and any number of other social and transactional applications you can think of.[25] Cumulatively, it provides Chinese citizens an easy method of managing almost all aspects of their lives. But it also provides state-controlled companies, local governments, and the Communist Party a means of looking into the details of individual citizens' lives that, by any democratic standard, would never be condoned.

Chinese bank accounts are linked directly to WeChat, and transactions are accomplished by direct, mobile pay (now using face verification). The users' transactional histories and banking details are accessible to the app provider . . . and, consequently, to the government itself. Even in the conduct of their offline daily activities, almost all purchases are made via phone. For purchases and payments of every kind—including food from grocery stores and restaurants, goods and products at retail stores and markets, bike and car rentals, entertainment and transportation tickets, monthly mortgages, apartment rental payments, utility bills, and even government taxes—payments are made the same way, usually by scanning square two-dimensional barcodes, commonly called QR codes.[26] All of it, transacted digitally and immediately traceable and forever recorded, makes the information regarding individual finances, spending habits, and financial status available and collectible by the systems that feed the government's calculation of "social trustworthiness" scores.

And the social credit scores of China's citizens aren't affected only by their online activities, offline purchase and payment histories, and apparent fiscal responsibility. Offline, nonfinancial behaviors—both personal and social—at home, at work, and in the community are also increasingly being tracked and calculated. Social offenses that diminish social credit scores include smoking in public places, playing music too loud, texting while driving, taking drugs, purchasing alcohol, being publicly intoxicated, arguing with spouses, spreading information considered false or unacceptable, espousing religious beliefs, loitering, littering, and even walking pets without leashes.[27] In essence, any violation of "acceptable" social norms can adversely impact one's social credit score.

As a result, Chinese citizens can find themselves blacklisted or otherwise restricted from renting cars, buying train or airplane tickets, obtaining favorable loan rates, acquiring insurance, purchasing real estate or otherwise obtaining affordable housing, making financial investments, and even attending preferred schools or qualifying for certain jobs and career opportunities.[28] Again, many Chinese consider these restraints fair exchanges for other government and technological benefits. It's important to realize, however, that most of China's population has never had the opportunity to enjoy anything approaching the sort of privacy rights generally expected and considered fundamental in democratic societies. As a consequence of its long history of authoritarian rule, many basic freedoms have never been granted in China as absolute entitlements. They've only been provided, when and if at all, from government authority and only in exchange for conformity or social silence, a quid pro quo. What citizens of democratic governments consider and demand as undeniable rights, Chinese citizens have only enjoyed as trade-offs for behaving consistently with Communist Party ideology. This is, after all, a government that effectively imposed a one-child policy on all married couples from 1979 until 2015—although, for some of those years, couples were allowed to have a second child, but only if their first was a girl.[29] In any event, there's little the Chinese can do to alter their reality. As has long been the Communist Party strategy, conformity is the only real option.

And just as technology is used to inform authorities of nonconforming behavior, it's also now being used to actively measure proof *of* conformity—in some cases by tracking citizens' consumption of loyalist Communist Party information and propaganda. In early 2019, a mobile app called *Xuexi Qiangguo* was released by the Communist Party's publicity arm that requires users to register by providing their mobile phone numbers and full names. Translated roughly as "study Xi to strengthen the nation," the app *allows* users to earn "study" points by logging onto it, reading articles, watching videos and documentaries about Xi Jinping, and taking multiple-choice quizzes on what they've learned about the party's policies and doctrines. Government directives were issued following the app's release that instructed party staff (or cadres) and members, of which there are 90 million, to download the app—with strong suggestions that they use it every day to increase their points, exhibit their loyalty, and earn benefits. *Xuexi Qiangguo* quickly became the most downloaded app in China, and yet another tool of state information efforts and social control. Before long, workers and users of the app found themselves spending long hours logged on, each day—feeling forced by the party, their employers, and colleagues to earn certain point levels. Reports surfaced that individuals' scores were being posted as another way to cause social shaming of those not showing or proving sufficient proof of party loyalty and communist ideals.[30]

————

While some contend that China's use of digital and AI technologies shouldn't be criticized—and that Xi's government is entitled to their applications as somehow culturally appropriate and politically acceptable—widespread reports of what's transpiring in China's largest western region argue otherwise.

Well over 90 percent of mainland China's population is composed of the Han Chinese. All Han share a deeply rooted, common genetic ancestry tracing back to ancient civilizations that originally inhabited a single region along the Yellow River in northern China. Throughout most of China's recorded history, the Han Chinese have been the culturally dominant majority. The Uighurs, on the other hand, are a minority of about 10 million people who live primarily in a large, autonomous region of China called Xinjiang.[31] The region came under Chinese rule in the eighteenth century and, geographically, has remained important to China for a variety of reasons—including that it's rich in natural resources and also that its western zone borders Tibet (another Chinese autonomous region) and eight foreign countries.[32]

Most Uighurs are Muslim. Due to their cultural differences from the Chinese majority, frictions with the Communist Party and central government have existed for many decades. The tensions increased dramatically, however, when separatist Uighur groups became more inspired in the 1990s by the collapse of the Soviet Union and by the growing number of independent Muslim states in Central Asia. The Uighurs began to demand their own cultural and political freedoms. But the Chinese government was quick to repress Uighur activism and, since then, hostilities have only increased. In recent years, there have been worldwide accusations and consensus that China is guilty of extreme human rights abuses against the Uighurs. It's now commonly reported that China is detaining between one and two million Uighurs, or 10 to 20 percent of all Uighur people, in more than a hundred detention camps in Xinjiang. Most of those detained aren't accused of any crimes, and very few records or information are even publicly available.[33]

For years, China denied the existence of the camps, but now *defends* their existence by claiming they are "vocational education centers," created only to stop the spread of religious extremism and terrorist activities, and also to "reeducate" the Uighurs by providing them with useful language, technical, and other skills. Loud critics of the Chinese policy, including many human rights organizations and the United Nations, say the practices are actually meant to quash the Uighur culture by conforming its population to Chinese norms and government social standards through isolation, deprivation, torture, and other extreme human rights violations.

In November 2019, an extraordinarily rare leak of over 400 documents from inside the Communist Party itself reached the *New York Times*. The documents confirm that the surveillance, control, and detention of the Uighur population in Xinjiang's reeducation camps are all occurring under the direct authority of Xi. Even more alarming, the leaked documents also expressly confirm that Uighurs are being taken into custody and detained without having committed any crimes. In fact, one document provides a literal "script" for local police and other officials to follow if asked by family members whether their missing relatives had committed any crimes. The official answer the script mandates is: "It is just that their thinking has been infected by unhealthy thought. Freedom is only possible when this 'virus' in their thinking is eradicated and they are in good health."

Also, consistent with the Communist Party and central government's approach elsewhere in the country, officials are using a scoring system to determine when, or if, those detained will be released. One document specifically instructs officials to tell inquiring family members that their *own* behavior could compromise their detained relatives' scores. Specifically, authorities are advised to say: "Family members, including you, must abide by the state's laws and rules and not believe or spread rumors. Only then can you add points for your family member, and after a period of assessment they can leave the school if they meet course completion standards."[34]

While China is receiving increasing international condemnation for its repression of the Uighurs, the atrocities have apparently continued unabated. In late 2019 and early 2020, the world learned that the Chinese government had systematically destroyed dozens of Uighur cemeteries in and around Xinjiang in an effort to eradicate their ethnic identity.[35]

The world is now also learning that Xi's government is beginning to use advanced facial recognition technologies to identify and track Uighurs not only in Xinjiang, but throughout other parts of the country as well. It is, to put it bluntly but accurately, the implementation of AI for government-funded, state-purposed racial profiling.[36] Uighurs, who descend from a different genetic lineage than the Han majority of Chinese, tend to have more Central Asian and Mediterranean facial features, a different skin tone, and a generally different physical build. Facial recognition software integrated into China's national surveillance cameras is now reportedly being used for "minority identification" purposes, specifically to spot and identify people of Uighur heritage. Through deep-learning algorithms that can identify patterns of facial features, databases are apparently being built by police forces and other agencies in various parts of the country to keep and analyze records of the whereabouts and activities of the Uighurs.[37]

We previously addressed the risks of inadvertent AI biases in Chapter 10. This bias, however, is different. It's intentional, purposeful, and designed specifically to implement racist distinctions for detection, decision-making, social policing, and ethnic control purposes. Like any machine learning program, the technology will only become more accurate and effective with the additional implementations and data to which it's exposed. As the algorithms of these applications become more adept at this type of use, the risk that they can be utilized or sold in other parts of the world becomes greater as well. It's not hard to imagine the implications of other countries obtaining these types of programs from the Chinese to bolster their own authoritarian or racially oriented models of governance.

———

Consistent with China's Belt and Road Initiative discussed in the last chapter, the nation is making its AI and related technologies available to influence other governments and regions it considers economically, politically, or militarily advantageous. Especially for countries unable to develop certain technologies or capabilities for themselves, Xi Jinping's willingness to export digital infrastructures and systems that can quickly enhance technological capabilities makes China a compellingly attractive economic and potential political partner.

In the South American state of Ecuador, thousands of cameras are now mounted on poles, buildings, and other structures throughout cities, villages, and much of the countryside. Purchased from China as part of a larger infusion of infrastructural support, the surveillance cameras were a part of contracts that provided Chinese financing, technology, materials, expertise, and training in exchange for payment obligations and rights to Ecuadorian oil reserves. The cameras, collectively referred to as ECU-911, are linked directly to dozens of monitoring stations where Ecuadorian police surveil the daily activities of citizens, ostensibly looking only for signs of crime. But it's become clear that the footage from those cameras is also accessed by Ecuador's domestic intelligence agency and that the surveillance system is routinely used for political repression and social intimidation purposes.[38] Although the cameras currently only provide video, audio, and phone-tracking capabilities, they're capable of supporting face-recognition software if, and presumably when, China agrees to provide it. China has sold similar surveillance systems to other countries as well, including Zimbabwe, Uzbekistan, Pakistan, Kenya, Venezuela, Bolivia, Angola, and the United Arab Emirates.[39]

Ecuador's ECU-911 system, and others like it, were manufactured and supported by two Chinese companies—the fully-state-owned China National Electronics Import & Export Corporation (CEIEC) and Huawei (pronounced "wow-way").[40] While CEIEC is an admitted, wholly-state-owned enterprise, Huawei's specific ownership is more of a mystery. Officially, the company claims to be 99 percent owned by its employees, their interests purportedly flowing indirectly through a labor union. To date, though, outside experts haven't been able to clarify the true structure.[41] What is clear, however, is that Huawei is linked to China's party-state in ways even more direct than those of most Chinese enterprises, and the government's intelligence agencies undoubtedly have leverage and influence over the company's decisions, activities, and data—a disconcerting reality given that Huawei has exported telecommunication infrastructures, equipment, and related consumer electronics to more than 150 countries around the world.

The next great change in digital technology and capability will come in the form of 5G technologies, in which Huawei is an industry leader. The term 5G stands for the "fifth generation" of wireless cellular technology, which won't just be an improvement over 3G and 4G capabilities, it will be a transformation. Engineered to operate using millimeter radio waves as signals, 5G networks will transform the internet with a broadband capacity perhaps 100 times the capacity of current 4G networks, and with network response times that will be 10 to 100 times faster than 4G. While a clean, uninterrupted connection to a 4G network produces response times of about 45 milliseconds, a 5G network will produce response times possibly less than 1 millisecond,[42] which is 300 to 400 times faster than the blink of an eye.

Because of the capacity and speed capabilities of 5G, it remains to be seen exactly what kinds of eventual applications will be possible, but it's clear that the system will enable digital transfers fast enough to provide seamless video, incredible download capabilities, proficient self-driving car technologies, and much more. Most significant, however, is the expectation that 5G speeds will enable AI applications to function at the individual mobile-user level—bringing the true future of machine learning technology, literally, into the hands of everyday consumers.

Upcoming 5G technology has entirely different infrastructure and hardware requirements than current 3G and 4G wireless technology. At the time of this writing, 5G is available in very few locations around the globe. But the transition from 4G is well underway, and, consistent with the Belt and Road Initiative, Huawei is aggressively marketing its ability to provide 5G core infrastructures and consumer devices to countries and regions throughout the world. Given the company's ties to China's party-state government, however, and its apparent legal obligations to the Chinese intelligence agencies, many countries see Huawei as a potential security threat and fear that any network reliant upon Huawei's 5G technology would be profoundly compromising—if not immediately, then at some point in the future.

For years, experts in the US and elsewhere have been concerned about the potential security breaches and attacks on critical infrastructure that could result from using Huawei telecom equipment, or even from sharing information with other countries that do. For security concerns, the US, Australia, New Zealand, Japan, and Taiwan have banned Huawei from participating in their 5G networks. In Europe, an outright ban of Huawei would be difficult to accomplish since, in the last decade alone, Huawei has become the supplier of a third of Europe's telecommunication systems. Nonetheless, the UK has elected to preclude Huawei equipment at the core, infrastructural level of its eventual 5G network, but it will allow Huawei to build out other aspects of the British system. Denmark's telecom provider is doing the same, and Germany and France have indicated they will increase security safeguards against any back doors that may be nefarious design elements of Huawei's technology. On the other hand, Russia, Czechoslovakia, Italy,

Belgium, and Brazil have all indicated that they will not limit Huawei technology. Numbers of other countries—including South Korea, the Philippines, Thailand, and other Southeast Asian nations—have already implemented Huawei technology in their 5G networks, and China claims it has 5G contracts with at least 25 other independent nations.[43]

In the US, the Department of Defense has banned sales of Huawei products on military bases, the Federal Communications Commission (FCC) has proposed rules that would formally prevent any American telecom company from using Huawei equipment, and various other legislation is being proposed to protect the country's infrastructure from the risks Huawei might cause. Also, since late 2018, the United States has openly tried to dissuade its allies from using Huawei as either an infrastructure or equipment provider, citing security concerns and compromises that could result from sharing intelligence or other information over its systems.[44]

———

Militarily, China doesn't approach the size, power, or sophistication of the US and its allies, which lead by a wide margin on the ground, in the air, and at sea. But China views AI technology as its opportunity to leapfrog certain phases of weapons development to bridge the gap between it and the US. In October 2018, the deputy director of the General Office of China's Central Military Commission confirmed the Chinese military's vision for AI when he characterized China's overall goal as an effort to "narrow the gap between the Chinese military and global advanced powers" by taking advantage of the "ongoing military revolution . . . centered on information technology and intelligent technology."[45]

That comment is completely consistent with all of China's other applications of AI that we've discussed. Despite that, China has unsurprisingly postured publicly that it wants to avoid an AI military race. In April 2018, China's ambassador to the United Nations submitted a position paper in which China claimed to support a worldwide ban on "lethal autonomous weapons." The paper, however, included definitions of lethal autonomous weapons that left conspicuously wide room for most actual and foreseeable forms of autonomous systems.[46]

In addition to the ambiguity of its public statements, it's also impossible to ignore the fact that China is unquestionably focused on the development of autonomous, unmanned military vehicles for deployment in ground, air, and underwater theaters of engagement. China has a long history of exporting arms to other countries, regardless of the recipients' political orientations, internal stabilities, international conflicts, United Nations status, or intended or actual use of the weapons systems.[47] There's no reason to believe Xi's administration won't continue to do so with AI-enhanced weapons systems, especially considering its intent to expand its sphere of technological and economic influence, as discussed throughout this and the previous chapter.

CHAPTER 14

RUSSIAN DISRUPTION

Pay no attention to that man behind the curtain.

—The Wizard of Oz *(1939)*
Screenplay by Noel Langley,
Florence Ryerson, and Edgar Allan Woolf

Just as the Chinese government's AI strategies are extensions of its political and ideological aspirations, the Russian government's uses of AI are equally consistent with the Kremlin's agendas.

The Union of Soviet Socialist Republics (the USSR or Soviet Union) began in 1922 as an association of previously independent republics that agreed to consolidate under a centralized, one-party system of government led by the Russian Communist Party based in Moscow. Following

World War II, the Soviet Union expanded to include territory covering one-sixth of the Earth's landmass. Until the final years of its downfall, it maintained a political, economic, and military superpower status that rivaled the US.

But when the Soviet system fell in 1991, the union dissolved back into 15 independent nations.[1] Russia, by far the strongest of them, remained the geographically largest nation on Earth. Still spanning 11 time zones and occupying 11 percent of the planet's land surface, it has a total area almost the size of the next two largest countries, Canada and the US, combined.[2] After the Soviet Union's disintegration, Russia reorganized politically, formed a new constitutional structure, and became the entity we know today. Now formally called the Russian Federation, it took over the Soviet Union's seat at the United Nations and became responsible under international law for all of the Soviet Union's prior global commitments.[3]

Throughout the 1990s, Russia struggled through political reformation and economic turmoil. Its economic output dropped by nearly 50 percent, and poverty rates hovered consistently above 35 percent.[4] But when Vladimir Putin became the Russian president in 2000, things began to change. During his years in office—a period that has now spanned the American presidential administrations of Clinton, Bush, Obama, and Trump—Putin has greatly reduced the country's poverty percentage, lowered its personal and corporate tax rates, increased wages, and enhanced the country's consumption and general standard of living. All of Putin's reformations have resulted in a middle class that's grown by tremendous numbers since he first took office.

As a former KGB agent and onetime head of the KGB's successor agency, the Russian Federal Security Service (FSB), Putin entered office with deep connections and alliances to the nation's controlling intelligence and military agencies. Although Russia is constitutionally structured as a multiparty democracy, under Putin's leadership it's in fact something far different. Better described as a bureaucratic autocracy, any political opposition that threatens Putin's standing is routinely suppressed, as are any unfavorable domestic press or media reports. And although Putin's government generally improved the standards of life for its common citizens, those improvements only came in exchange for Putin's privatization of the country's true wealth into the hands of a small group of ultra-rich oligarchs. Many of them followed and supported Putin into power from his days with the KGB and FSB. Although the laws of most democracies prevent elected officials from enriching themselves or others by unfairly directing government contracts or preferential treatment to personally preferred businesses, Putin has used his office and influence to assist approximately 100 of his supporting oligarchs in acquiring more than a third of the country's overall wealth.[5] His own finances have benefited during his time in office as well. Although he downplays his financial holdings at every opportunity, many experts believe Putin has become one of the world's wealthiest individuals since taking power—with a fortune spread across a wide range of secretly held Russian oil, natural gas, real estate, and other corporate interests.[6]

While he and a small number of individuals have become extremely rich during Putin's governance, the country itself is still struggling to reestablish its own economic health. From 2014 to 2017, the Russian economy fell into an especially deep state of crisis caused by dramatic drops in oil prices (its major export) and strict economic sanctions imposed by the US, the European Union, and other nations after Russia encroached eastern Ukraine and annexed Crimea.[7]

Those were of course critical years in the development of AI due to the emergence of machine learning. As mentioned in the opening prologue to this book, Putin made no qualms about telling the world in September 2017, "Whoever becomes the leader in (AI) will become the ruler of the world."[8] He has an inherent problem, however, in making his own country that leader. As a practical matter, the Russian government's approach to AI is significantly constrained by the realities of the country's long-struggling economy, an oligarch-controlled business environment, and a notorious reputation for widespread corruption. Together, those factors have nurtured a poor atmosphere for outside investors and enterprises. As a consequence of that background and the ongoing international trade and related difficulties confronting Russia, the country is dramatically far behind both the US and China in AI investments, research facilities, expert talent, and development capabilities. Simply put, Russia doesn't have the available funding—from either domestic or foreign sources—and is without the technological infrastructure and expertise required to match the level of AI efforts and accomplishments taking place in other parts of the world.

That's not to say, however, that the Russians aren't fully committed to making AI work to their international advantage. They are. But the paths they're following have relatively singular end lines in sight. In October 2019, Russia released a long-awaited national AI strategy. Originally authored at Putin's request by Russia's largest bank, Sberbank, the strategy generally focuses on technological security, transparency, sovereignty, innovation integrity, cost-effectiveness, and support for competition. The plan envisions that, over a ten-year period, Russia will increase its scientific research and development efforts; invest in software and hardware; enhance the availability and quality of its data; improve its ability to educate, retain, and attract top-quality AI talent; create and maintain a favorable and flexible AI regulatory environment; and integrate AI-based technologies into various sectors of the Russian economy and society—most notably in its national healthcare sector.[9]

Conspicuously absent from the plan are specific budgetary commitments or any mention of AI development for national security, defense, or military purposes. Regardless of those omissions, the Russians undoubtedly recognize they are far behind and economically incapable of challenging the US or China in overall AI research or development. The Kremlin is therefore expected to dedicate a high percentage of its limited resources to pursue competitive AI advantages in two distinct and very narrowly defined lanes: military weapons technology and cyber/internet disinformation strategies.

With respect to AI-enabled military systems, Russian software engineers don't have access to the technological ecosystems of their Western counterparts. In today's world, however, software is readily available on the open market, at least at core, foundational levels. Many of the world's companies that develop AI for modern video games, like the battle games we discussed in Chapter 7 for instance, routinely make their platforms openly available for all to access—the theory being that AI should be open-sourced and the coding transparent. Replicating and modifying those algorithms to perform in real-world situations, and on real-world military planning software, is not an overly difficult task to achieve—even for comparatively lesser-equipped engineers. Similarly, the computer vision algorithms that drive many facial recognition technologies around the world are also obtainable by weapon developers capable of modifying them for actual, live-theater military purposes.

Realizing their economic obstacles, Putin's defense agencies and military designers are aggressively putting machine learning technologies to their most immediately accomplishable and impactful uses—electronic warfare (EW) and robotic weapons. In 2017, Russia deployed EW units to Syria, eastern Ukraine, and Crimea, where they were used to gather data about the performance capabilities and electronic signature characteristics of American and other Western aircraft, naval vessels, and missile systems in the region.[10] Russia has also developed and deployed a number of semi- and fully autonomous battlefield robotic systems, including manned and unmanned ground, air, and naval vehicles that are either AI enhanced or fully AI controlled for combat, intelligence gathering, and logistical support roles.[11] Putin has even boasted that Syria was a "priceless" opportunity for them to test more than 200 new weapons.[12] At this point it's believed they've acquired an entire roster of weapon systems—including tanks, planes, submarines, and swarm missile systems—that incorporate neural networks and machine learning algorithms that can autonomously identify and select perceived enemy targets and then "decide" for themselves whether to engage those targets.[13] Lethal autonomous weapons systems (LAWS) like these are the subject of significant worldwide debate, including ongoing policy considerations at the United Nations.[14]

The second AI track the Kremlin is focused on is domestic and international propaganda, surveillance, and disinformation. Since Putin first became president, mandates to control and manipulate information have been key components of his policies. Now, some 20 years later, Putin's administration is still intent on accomplishing its own form of domestic digital authoritarianism. The government's control of traditional and digital media sources and its repression of independent media outlets have increased under Putin's reign. There are more reporters in Russian prisons now than at any point since the fall of Soviet Russia.[15] Digital surveillance and social control strategies have been enhanced. Russian social and political speech is monitored carefully, especially for those considered activists or political adversaries, and Putin is now looking to create an independent, sovereign internet that will be fully controlled by the Kremlin and shield all of Russia from vast amounts of outside information,[16] akin to the Great Firewall of China.

Russia's System of Operative Search Measures (SORM) was first created in 1995 and requires all Russian telecommunications and internet providers to install hardware provided by the FSB that gives it the ability to monitor Russian phone calls, emails, texts, and web browsing activities.[17] Five years later, during Putin's first week in office, he expanded the SORM's reach by allowing a number of additional Russian security agencies apart from the FSB to gather SORM information from Russian citizens and foreign visitors. The agencies that can now legally collect phone and digital data as a matter of course from anyone within the borders of Russia include the Russian federal police, the Interior Ministry, the Federal Protective Service, the Foreign Intelligence Service, Russian customs, the Federal Drug Control Service, the Federal Penitentiary Service, and the Main (Intelligence) Directorate of the General Staff.[18] The government now also employs sophisticated facial recognition technologies on camera and other systems throughout the country, particularly in Russia's major cities of Moscow and St. Petersburg, to track the activities of individuals it chooses to target.[19]

Outside of its borders, Russia may not have the degree of direct international economic and other influence it would like, but Putin's government and military are proving themselves quite capable of influencing countries in other ways. It is a standard doctrine of Russian military strategy to conduct information warfare (*"informatsionaya voyna"*) to interfere in the politics and operations of its foreign adversaries through cyber and other operations. The 2010 "Military Doctrine of the Russian Federation" specifically says such measures are taken, "to achieve political objectives without the utilization of military force."[20]

This is nothing new. Information warfare is a long-held Russian military concept that goes back to the earliest days of the Cold War. As General Valery Gerasimov, the chief of the general staff of the armed forces of Russia, publicly acknowledged as recently as March 2019, the Russian government and military consider it a simple reality of international power and politics that they should, and do, conduct information and propaganda campaigns, including political interference, as an integral part of their regular national defense strategies.[21] Even the most recently published "Military Doctrine of the Russian Federation" (2015) expressly states that one feature of modern military conflict is "exerting simultaneous pressure on the enemy throughout the enemy's territory in the global information space."[22] Further, and perhaps most pertinent, Russian military doctrine does not differentiate between times of war and times of peace with respect to strategic noncombat measures waged against adversaries . . . and, by any objective and informed account, Russia considers *any* country of significant global standing that is not its formal ally to be its adversary.

Russian information warfare tactics don't have the absolute goal of convincing foreign populations that disinformation and lies are necessarily the truth. Instead, Putin's Russia considers it strategically sufficient just to plant seeds of confusion, doubt, and disruption in the populations of foreign adversaries. The goal, first and foremost, is to internally polarize populations. Leading political theorists have long recognized disinformation as a basic tenet of governments with total-itarian orientations. Twentieth-century German American philosopher Hannah Arendt said it well: "A people that no longer can believe any-thing cannot make up its mind. It is deprived not only of its capacity to act but also of its capacity to think and to judge. And with such a people you can do what you please."[23] That axiom holds true in both domestic and foreign policy, and the creation and distribution of disinformation has been a Russian political methodology predating the Soviets. Under Putin, *digital* disinformation has become a vital component of Russian foreign policy strategies. To Russian cyber operatives, fake news stories implanted within foreign social media and news platforms are consid-ered doctrinally appropriate political and military measures. They reg-ularly use AI-enabled algorithms to create and manipulate information to spread provocative and intentionally misleading political reports, social slogans, fake news stories, and contrived or mislabeled photos and videos.

With respect to the latter, in 2019, an entirely new and dangerous category of AI disinformation technology began to emerge called *deep-fakes*. Using machine learning techniques, a deepfake is a video and/or audio clip that shows individuals appearing to do or say things that, in actuality, were never done or said—essentially creating events that never occurred. The danger of such technology can't be overstated, and its potential to sow discord by adversely affecting public impression, opinion, and politics is significant.

Russian meddling in foreign elections has been occurring for decades, and recent intrusions by Russian operatives into European and Amer-ican national elections are open examples of their disinformation and interference strategies and capabilities.[24] The Russian efforts to manipu-late the American 2016 presidential election are well-known. Through-out the entirety of the US intelligence network, it is an accepted fact that Russian citizens working out of a cyber-troll factory in St. Peters-burg, known as the Internet Research Agency (IRA), set up various fake social media accounts through which the kinds of AI-enabled internet bots described in Chapter 11 were used to spread a variety of contrived content aimed at polarizing the American electorate and alienating cultural, social, and political groups from one another.[25]

There's no question that the IRA operated with the knowledge and support of Russian intelligence services.[26] Following the 2016 American national election, the US attorney general created a special counsel's office to investigate Russian interference, including any links or coordination between the Russian government and individuals associated with the Trump campaign. In the special counsel's own words on May 29, 2019, he summarized his team's conclusions as follows:

> Russian intelligence officers who were part of the Russian military launched a concerted attack on our political system.
>
> . . . [T]hey used sophisticated cyber techniques to hack into computers and networks used by the Clinton campaign. They stole private information, and then released that information through fake online identities and through the organization WikiLeaks. The releases were designed and timed to interfere with our election and to damage a presidential candidate.
>
> And at the same time . . . a private Russian entity engaged in a social media operation where Russian citizens posed as Americans in order to interfere in the election.
>
> . . .
>
> I will close by reiterating . . . that there were multiple, systematic efforts to interfere in our election.[27]

To what extent the Russian efforts materially influenced the actual outcome of the 2016 US election is, for purposes of this conversation, irrelevant. What should alarm every American citizen, in fact every world citizen, is that intelligence agencies across the globe had little doubt that Russia would continue those interference strategies in the future. Putin's denials of the Kremlin's actions and intentions—past, present, or future—shouldn't, when taken on their own, merit unbridled trust. Who would expect him to publicly admit to his government's involvement? And, in any event, why should his denials matter when Russian military doctrine makes their strategies and intentions clear? As AI disinformation applications grow more sophisticated, it's only reasonable to expect that Russian interference efforts and threats will as well. And, as predicted, the Russians continued their efforts throughout the run-up to the 2020 American general election as well. These are matters not only the US, but all of the Western world, must address and purposefully defend against going forward. Technology and tactics in the disinformation realm will always evolve and will always be a cause for concern. The goal for the West should be to actively anticipate Russian strategies . . . and to defend from positions of coordinated *advantage*, rather than latent response.

In summary, Russia's economic and infrastructure shortcomings prevent it from becoming a leading developer of AI and consequently compromise its ability to pursue AI technologies with any hope of directly acquiring significant economic reward. Russia is therefore following AI strategies that, from its perspective, otherwise generate value. Under Putin's governance, Russian agencies and operatives are, and will continue, using AI as a new tool to exert overt military pressure on their adversaries and impose covert cultural and political disruption beyond their borders. Those strategies have been, from the time of the Soviet Union, the Russian government's way.

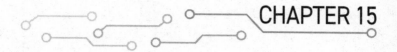

CHAPTER 15

DEMOCRATIC IDEALS
IN AN AI WORLD

Democracy is not so much a form of government as a set of principles.

—Woodrow Wilson, 1856–1924
Twenty-Eighth President of the United States

Democratic systems of government first emerged in ancient Athens when Cleisthenes, who is generally regarded as the father of Athenian democracy, proposed changes to Athens's constitution that would end centuries of past leadership by aristocratic and tyrannical rulers who put their own interests above those of the citizenry. Cleisthenes's new form of government gave an equal vote to all male citizens regardless of class, so long as they weren't slaves.[1] That was about 2,500 years ago, around 500 BCE. By today's standards, it was a crude and incomplete beginning, but it set in motion the earliest foundation of a concept that still underlies democracy today—that individual citizens should have an acknowledged and equal voice in choosing their government leaders and in deciding matters related to their individual rights. Other Greek cities developed democratic variants of the Athenian model, but none were as successful or well structured. Despite periods of disruption along the way, democracy in Athens lasted almost 200 years—which was a remarkable run considering the unstable nature of the times. In the end, the world's first democracy succumbed to internal conflicts, outside invasions and, eventually, conquest by the Macedonians. By 323 BCE, most aspects of Athenian democracy had disappeared.[2]

After the demise of Athens, it took almost 2,000 years for any significant resurrection of democratic principles in a major national government. Despite periodic spatterings of systems that allowed elections and limited voting rights in various parts of the world during the Middle Ages, it took until the late sixteenth century for the people's right to govern themselves—or at least to have a legally mandated say in how they were governed—to reemerge. In the early 1600s, the English Parliament became increasingly unhappy with the absolute rule of the British monarchy. Inspired by an ancient English document, the Magna Carta, which King John had signed in 1215 to make peace with rebellious land barons, the notion that English citizens were *legally* entitled to certain rights and freedoms gained gradual momentum. In 1628, Parliament passed the Petition of Right, which was a major step toward restricting the absolute power of the monarchy by seeking certain liberties for English citizens, including no taxation without the consent of Parliament, no imprisonment without cause, no required quartering of soldiers by subjects, and no martial law in peacetime.[3]

Although King Charles I eventually accepted Parliament's petition, he ultimately ignored its rules. Fifty years later, however, and after a period of restoration that again strengthened autocratic rule in England, Parliament in 1679 passed the Habeas Corpus Act, which prevented the detention of citizens without sufficient cause or evidence. In 1689, the seminal English Bill of Rights was enacted. It limited the powers of the monarch, set out the rights of Parliament, established regular and free elections, and codified certain individual rights—including freedom of speech within Parliament, the prohibition of cruel and unusual punishment, and Protestant citizens' right to bear arms for their own defense. It also established, once and for all, that English citizens would be free of any taxation without Parliament's agreement.[4] For the first time, the absolute power of the monarch was put in check.

The *most* significant step in the advance of modern democracy, however, didn't occur for many more years. It again involved England and its parliament, but took place an ocean away. In the mid-1700s, the combined population of the 13 British colonies in America had grown to well over a million people,[5] all of whom were subject to the rule of Great Britain even though they lacked formal representation in Parliament. By the 1760s, Britain was increasingly imposing oppressive and restrictive taxes on the colonies for the sole benefit of the Crown's own treasury. The travesty of the situation wasn't lost on the colonists—taxation without representation had been an oppression that English subjects had opposed for centuries. Then, in 1765, Parliament passed another tax the American colonists considered a final, unbearable weight. Called the Stamp Act,[6] it required that almost all documents printed in the Americas use paper produced only in London. It was an outright effort to fully control the only medium through which most documented communications and formal transactions could be made, as it would have applied to all forms of pamphlets, newspapers, books, licenses, legal documents, shipping logs, and ledgers.[7] *Consider the impact: in an analog age, Britain's intent was to restrict all colonial written transactions and records to a platform imposed upon the colonies from outside their cultural borders. In today's digital atmosphere, China's aspirations to spread its 5G infrastructure to other nations who lack available alternatives, and who will then be functionally and economically dependent upon a foreign entity, is not entirely different.*

By the early 1770s, the growing tensions between the British and Americans had become violent. In combination, the Boston Massacre, the Boston Tea Party, additional confrontations between the colonists and British soldiers, and Britain's imposition of retaliatory taxes and port embargoes all led, in April 1775, to the confrontation of military forces at the Battles of Lexington and Concord and to the formal outbreak of the American Revolutionary War.

A year into the conflict, congressional representatives from the 13 colonies met to draft a unified statement justifying their intent to fully separate from British sovereignty and rule. The document they created was the American Declaration of Independence. Thomas Jefferson, who represented the colony of Virginia, wrote the initial draft and is generally considered the Declaration's principal author. The final version of his document was signed by the American Congress on July 4, 1776, and from that date forward, the colonies considered themselves independent sovereign states no longer subject to British rule.[8]

The underlying and primary spirit behind Jefferson's Declaration of Independence was eloquently stated in its second sentence. As one of the best-known sentences in the English language, and arguably the most powerful single statement of the moral ground upon which the United States' form of democracy is structured, it reads:

> We hold these truths to be self-evident, that all men are created equal, that they are endowed by their Creator with certain unalienable Rights, that among these are Life, Liberty and the pursuit of Happiness.[9]

Despite the colonists' formal Declaration of Independence, the Revolutionary War raged on for an additional seven years. More than 200 separate battles were fought, with thousands of casualties on both sides.[10] The Treaty of Paris finally ended the revolution in 1783, formalizing America's independence and national sovereignty.[11] Six years later, the United States Constitution became effective. It established a federal system of government designed to recognize and ensure—first, foremost, and for the future—the individual personal rights and freedoms of its citizens. It established three branches of government, the separation of powers between them, and the separation of powers between the federal and individual state governments.[12] Two years later, in 1791, the American Bill of Rights was added, specifically guaranteeing the moral freedoms and personal protections that Jefferson first expressed in 1776.[13]

———

Just over half the world now has systems of government that can fairly be characterized as democratic, but the proliferation of democracy, even with the US as a principal model, has only occurred in the last 75 years. In 1945, at the close of World War II, there were only 12 democratic governments. Now, approximately 100 of the 195 states recognized by the United Nations are democracies in structure and overall ideology.[14] There are many types of democratic constructs, and the amount of actual power available to voters varies widely. Direct democracies allow the populace to vote on all significant issues through referendums called when needed. Representative democracies elect officials to represent the interests of their constituents when enacting law. Parliamentary democracies enable the party with the greatest representation to choose the highest government leaders and control legislation. Most systems, however, are hybrids that incorporate elements of each, and that often operate differently on local, regional, and national levels. The foundation in all democratic systems, though, is individual dignity—and the characteristic pillars of any government system that truly supports democratic ideals are free speech, free press, the right to an equality of vote, and a system of laws that not only supports those rights but that's also open to debate and revision as circumstances change or demand.

Vital to all forms of democracy is the citizenry's right to accurate information and the ability to become educated so they can meaningfully understand and express their interests, either directly or through their elected representatives. Britain's Sir Winston Churchill is often quoted as saying, "The best argument against democracy is a five-minute conversation with the average voter."[15] To put those words in the context of today's digital information age, an uninformed, misinformed, or disinformed electorate is no real electorate at all—as an unaware or confused voting base compromises the most critical component in the basic philosophy upon which democracy rests.

So, what does all of this mean in regard to the many new issues that artificial intelligence is bringing to the world, and with respect to the ways AI can be used by those who have no regard for the creation or preservation of individual rights, either within or outside their own country or culture? The answers are numerous, but by now many of them should be apparent. It means that the democracies of the world need to act both independently and in concert to ensure to the extent possible that their citizens enjoy the benefits of AI, but that they also remain protected from its detriments. Like all technology, AI is a dual-use tool. And as we've shown repeatedly throughout the prior pages, the purposes for which it is used can potentially bring great benefits, but also significant compromise. The moment is now for the West—particularly those nations in a position to significantly affect the development and deployment of AI—to openly engage its populations and policy makers in practical, transparent, and accountable conversations about AI and its related technologies and implications (including changes in workforce requirements), and to then impose agreements, regulations, and laws accordingly. This is another Sputnik moment. And this *must* be another point when purpose outpaces politics.

THE UNITED STATES

Like all democracies, actionable policy under the American system is at times complicated to accomplish, particularly when there's partisan disagreement—which there generally is. But, at the end of the process, the hope is that formal policies are never inconsistent with the American Constitution and with the promises of individual rights and liberties upon which it is structured. At the very least, it's imperative that regulations and laws reflect the informed and majority will of the people. Although the country and its Constitution were birthed some 250 years ago, the circumstances that first motivated America's formation still cast a very clear light on what the US approach to artificial intelligence must be.

While America's Founding Fathers couldn't possibly have imagined AI or the far-reaching nature of its implications, the constitutional structure they crafted nonetheless mandates the path American politicians, bureaucrats, and business leaders are obligated to follow in the days and years ahead. And while individual privacy rights aren't specifically addressed in the US Constitution or Bill of Rights, the nation's courts have consistently held that they both imply a variety of legally protected privacies.[16] State and federal laws now extend those rights of privacy to digital technology, making—in most cases—digital information entitled to the same protections as any other form of information or records.

In October 2016, the Obama administration took the first broad step toward American regulation of AI when it published a report by the National Science and Technology Council entitled "The National Artificial Intelligence Research and Development Strategic Plan."[17] When it was released, the country was in the final weeks of a particularly divisive presidential race and the plan consequently received little media recognition outside of technology circles. For the most part, mainstream American media remained focused on political dramas and distractions during the years following the 2016 election, causing AI—along with numbers of other globally impactful news matters—to remain conspicuously absent from everyday discourse.

Despite the lack of attention, the Obama national AI plan was a pragmatic and early description of the focused directions the US ought to pursue. It made expressly clear that narrow, task-specific applications of AI present tremendous economic and societal implications that need to be a primary focus. At the time, prominent philosophers, authors, and organizations around the world—including the Future of Life Institute at MIT,[18] the Future of Humanity Institute at the University of

Oxford,[19] and the Machine Intelligence Research Institute at the University of California, Berkeley[20]—were aggressively debating the risks of artificial *general* intelligence (AGI) and extolling the need for restrictive regulations to forestall or at least safely prepare for AGI's potential emergence. The Obama administration, however, took the contrary position that the American government's policies should be directly focused on the already-existing realities of narrow AI, and not motivated by either aspiring to achieve AGI or preparing for the hypothetical risks then commonly associated with it.

The positions expressed in the Obama plan were that narrow AI is an existing commercial reality, that general AI may never occur (at least not in the near or foreseeable future), and that, as a practical matter, the best way to prepare for any possible future emergence of AGI is to confront the risks immediately presented by narrow AI—such as security, privacy, and safety. The plan prioritized seven administration strategies for federally funded AI research, both inside and outside of government: (1) to make long-term investments in AI; (2) to develop effective methods for human-AI collaboration—rather than focusing on the fear of AI replacing humans; (3) to understand and address the ethical, legal, and societal implications of AI and to ensure its design in ways consistent with democratic norms; (4) to ensure the safety of AI systems by assuring its applications are reliable, dependable, and trustworthy in controlled and well-defined test environments; (5) to develop high-quality, accurate, and publicly shared data sets for AI training and testing; (6) to develop a broad spectrum of standards, benchmarks, and evaluative techniques to measure the veracity of AI; and (7) to determine the manpower needs of the American research community to ensure that a sufficient national workforce is available for the demands of future AI research and development.[21]

Even in a divided democratic country, it's clear that AI must be a bipartisan issue. Following Obama's departure from office, the Trump administration hosted a one-day meeting on artificial intelligence in May 2018, at which 100 senior government officials met with academic experts and American business leaders. After the event, the administration published a report entitled "Summary of the 2018 White House Summit on Artificial Intelligence for American Industry," along with a White House fact sheet, "Artificial Intelligence for the American People." In combination, the two documents highlighted the administration's general intent to: (1) prioritize funding for AI research and development for computing infrastructure, machine learning, and autonomous systems; (2) remove regulatory barriers to AI innovation; (3) prioritize science, technology, engineering, and math (STEM) education; (4) recognize the need for the US military to lead in AI technology; (5) call for government services to improve their efficiencies by using automation software and maximize the sharing of data with the American public; and (6) work with US allies to recognize the potential benefits of AI and to promote AI research and development.[22]

The Trump administration also established a cabinet-level Select Committee on Artificial Intelligence to advise and assist the National Science and Technology Council on national policy matters by providing a formal mechanism for interagency coordination and the development of federal artificial intelligence activities—including those related to autonomous systems, biometric identification, computer vision, human-computer interactions, machine learning, natural language processing, and robotics.[23] In addition, a National Security Commission on Artificial Intelligence was created to determine the methods necessary to develop AI, machine learning, and related technologies necessary to satisfy US national security needs, to assess the impact of AI on US competitiveness and technological advantages, to attract and recruit leading AI talent, and to assess the risks associated with advances in foreign military deployments of AI.[24] In January 2019, it was announced that executives from major US technology firms including Google, Microsoft, Amazon, and Oracle would serve on the commission, along with former national security officials and leading academics.[25]

In February 2019, President Trump signed a presidential executive order captioned "Maintaining American Leadership in Artificial

Intelligence."[26] Section 1 of the order sets out its purposes, and very well summarizes the American government's overall consensus approach to AI technologies.

(a) *Policy and Principles.* Artificial Intelligence (AI) promises to drive growth of the United States economy, enhance our economic and national security, and improve our quality of life. The United States is the world leader in AI research and development (R&D) and deployment. Continued American leadership in AI is of paramount importance to maintaining the economic and national security of the United States and to shaping the global evolution of AI in a manner consistent with our Nation's values, policies, and priorities. The Federal Government plays an important role in facilitating AI R&D, promoting the trust of the American people in the development and deployment of AI-related technologies, training a workforce capable of using AI in their occupations, and protecting the American AI technology base from attempted acquisition by strategic competitors and adversarial nations. Maintaining American leadership in AI requires a concerted effort to promote advancements in technology and innovation, while protecting American technology, economic and national security, civil liberties, privacy, and American values and enhancing international and industry collaboration with foreign partners and allies. It is the policy of the United States Government to sustain and enhance the scientific, technological, and economic leadership position of the United States in AI R&D and deployment through a coordinated Federal Government strategy . . .[27]

The executive order goes on to define various government agency objectives, the different roles and responsibilities of the agencies, and the need for the government to collaborate with industry, academia, international partners, allies, and other nonfederal entities.

The US Department of Defense (DoD) operates under the authority of the executive branch and is responsible for coordinating and supervising all agencies of the federal government directly concerned with national security and the armed forces. Its stated mission is to "provide the military forces needed to deter war and to protect the security" of the country.[28] With more than 2.8 million employees, the American DoD is the largest employer in the world—including 1.3 million active-duty servicemen and women, approximately 800,000 members of the National Guard and Reserves, and more than 740,000 civilian employees.[29] The DoD will itself be a significant developer and user of AI technologies in years going forward. As the mandated national defender of American rights and dignities, it will also be the country's primary protector in the face of foreign AI. In response to the global technological changes so rapidly occurring, along with the world's apparent return to an era of aggressive, strategic competition, the DoD is now taking meaningful steps to ensure the ethical design and use of AI, both domestically and abroad.

In February 2019, the DoD released a comprehensive artificial intelligence strategy that sets out the strategic approaches it will pursue to incorporate AI.[30] Most notably, the strategy calls for a reinvigorated and transparent partnership with the nation's private sector. As the DoD has on many occasions before, it again realizes that the current state of AI is yet another critical inflection point in the history of America that requires a cooperative synergy with the nation's corporate enterprises. Just as in the post-Sputnik years, all segments of the country must work together to ensure that American values and visions are not only accomplished but protected.

It would be incomplete to talk about AI activities in the US without also addressing the independent roles and responsibilities of the private sector. Unlike in China and Russia, American democratic capitalism allows for free enterprise to operate and evolve without the intrusive involvement of government. So long as companies and corporations abide by the law, they are entitled to produce products and services that are not specifically controlled by the state. More importantly, the intellectual property of those companies is not open to unbridled government access. Silicon Valley is the ultimate example of the heights American industry can reach when allowed to operate on its own. The technological innovations that have come from Silicon Valley have pushed the world forward as profoundly as the Industrial Revolution of the early 1800s. Silicon Valley's success was of its own making, spurred on by some of the greatest modern minds and entrepreneurs who were appropriately left to flourish in a freely competitive environment.

With that said, moments occur when industry is ethically challenged by outside influences that are inconsistent with the underlying ideals of democracy. An interesting example involves Saudi Arabia's Silicon Valley investment efforts. As a widely known national investment strategy—primarily to diversify away from its historical reliance on oil revenues—Saudi Arabia is now dedicating huge sums to digital technology and AI.

But in October 2018, news of Saudi Arabian dissident and American journalist Jamal Khashoggi's death made world headlines and caused an interesting reaction from private technology and AI enterprises. Khashoggi's death was coincidentally less than a month prior to the annual Future Investment Initiative (FII)—a major international forum hosted by the Saudis to discuss trends in the world economy and investment environment. Khashoggi had been a public critic of the Saudi royal family, and clear evidence quickly emerged that linked his brutal demise at the Saudi consulate in Istanbul, Turkey, to agents of the Saudi government. Although intelligence agencies from around the globe became convinced that the Saudi crown prince, Mohammad Bin Salman (MBS), had direct knowledge and had at least implicitly authorized Khashoggi's killing, he denied any involvement, either indirect or otherwise. Despite the denials, MBS and the Saudi royal family suffered serious criticism from outside the country. Khashoggi's death and its aftermath reminded many in the Western world of Saudi ruthlessness, and that much of what the Saudi government appears to stand for is antithetical to democratic ideals of human dignities and freedoms—particularly free speech and free press. The 2018 FII conference went forward, but many individual and corporate participants, scheduled speakers, government representatives, and media members boycotted the event.[31]

Another consequence of Khashoggi's death was that it gave technology start-ups and companies around the world reason to pause and carefully consider whether, morally or businesswise, they ought to accept financial investments and lucrative funding offers from Saudi Arabia.[32] In the US, foreign investments in Silicon Valley enterprises are becoming increasingly subject to federal government review and approval, primarily to ensure that Chinese investors with ties to Beijing aren't obtaining access to private American intellectual property.[33] While few suspect Saudi Arabia to have ever been guilty of corporate espionage, the Khashoggi event and the royal family's later indignant denials of any involvement nonetheless brought a level of open discussion about the propriety of partnering with Saudi money. Many Silicon Valley companies and boards of directors that hadn't previously thought twice reportedly began to consider whether they should carefully assess the moral and cultural standards of a nation that now considers data its new oil, AI its next great revenue stream, and American technology companies its next lucrative opportunity for huge investment. In 2016 alone, Saudi economic infusions in Silicon Valley AI ventures reportedly reached $5.3 billion.[34] While it's unclear whether any companies ultimately chose to decline significant offers of Saudi investments following Khashoggi's death, the topic was at least debated.

The Saudis also captured significant media attention when, at the same 2018 FII conference, they became the first country to grant national citizenship to a robot.[35] Developed by Hong Kong–based Hanson Robotics, Sophia is a machine learning robot enabled by voice recognition and visual interactive software. It's constructed to look like a human, with an artificial face modeled from actress Audrey Hepburn.[36] Though very impressive from engineering and robotics perspectives, from an AI perspective Sophia is arguably little more than a sophisticated chatbot—albeit one that's installed in a mechanical robot equipped with a complex array of motors and gears that enable it to seemingly "emote" a range of physical facial expressions.

The travesty apparently lost on the Saudis when they decided to run the media ploy of granting Sophia a Saudi Arabian citizenship is that, to most observant people, the publicity stunt did little more than highlight the secondary citizenship status of actual flesh-and-blood Saudi Arabian women.[37] It was stagecraft, shortsighted, and hypocritical. In the real world of Saudi Arabian culture, women still enjoy comparatively few legal freedoms or rights. They must have an official male guardian— who is usually a father, uncle, brother, or husband—and must have that guardian's consent to obtain a passport, travel, get married, get divorced, or sign contracts.[38] It's telling that, in any legal proceeding, a Saudi woman's testimony is only half as valid as a man's. Simply put, Saudi Arabian women are treated profoundly inferior to men.

On a separate, but directly related note, a Saudi mobile app available from both Apple and Google merits discussion. "Absher" (roughly translated as "Yes, Sir" or "Yes, done") is a product of the Saudi Interior Ministry that gives Saudi Arabian men the ability to exercise their guardian rights over women by tracking their locations and blocking their ability to travel, conduct financial transactions, and even obtain certain medical procedures.[39] To a country's leadership that considers it culturally appropriate and legally acceptable to discriminate and control the rights of women, this type of app is a perfectly acceptable and socially efficient tool of AI. As is clear from earlier pages, it should come as no surprise that countries and cultures—in fact, any country or culture—will use AI in ways they deem morally and legally acceptable. While the Absher app is but one example, it highlights an imperative question for private enterprise that develops AI under the freedoms provided by democratic principles. That question is whether companies should participate in or enable oppressive uses of their commercial technologies by countries with vastly contrary cultural and moral codes. These types of issues deserve transparent debate, and a cooperative and consistent approach from democratic governments and their private institutions alike.

THE EUROPEAN UNION

Many European countries share a united sense of respect for democratic ideals and individual human rights that are now beginning to fundamentally inform and direct their domestic and international policies with regard to AI.

In the aftermath of World War II, European governments understood all too well the societal devastations that could result from military conflict between any two or more of them. As countries across the European continent went about rebuilding their infrastructures and economies, they therefore concentrated also on strengthening and solidifying relations with their regional neighbors. In 1957, six countries—Belgium, West Germany, France, Italy, Luxembourg, and the Netherlands—entered into a formal agreement establishing the European Economic Community (EEC).[40] The primary concept behind the EEC was the idea that with an agreed system of economic reliance between the member countries, any one of them would be less likely drawn into a military conflict that involved or affected the others. The UK, Ireland, and Denmark joined the EEC in 1973,[41] and other countries became members over the years and decades that followed. In 1993, the EEC was renamed the European Union (EU), at which point it specifically mandated that any country wishing to join or remain a member must have a stable form of government that guarantees "democracy, the rule of law, human rights and respect for and protection of minorities, the existence of a functioning market economy as well as the capacity to cope with competitive pressure and market forces within the Union."[42] By 2013, the EU had grown to represent a combined population of more than 500 million Europeans from 28 countries:

Austria	Italy
Belgium	Latvia
Bulgaria	Lithuania
Croatia	Luxembourg
Cyprus	Malta
Czech Republic	Netherlands
Denmark	Poland
Estonia	Portugal
Finland	Romania
France	Slovakia
Germany	Slovenia
Greece	Spain
Hungary	Sweden
Ireland	United Kingdom

Over the decades of its operation, the economic and other benefits of participating in the EU became crucial, almost indispensable, to its members—as evidenced by the UK's long and nationally divisive Brexit debates about whether—and, if so, how—to withdraw from the institution, which it eventually did in early 2020.

Structurally, the EU is itself organized as a representative democracy, without a single leader, but instead with seven main legislative bodies with separate powers and interdependent areas of responsibility. The legislative bodies work cooperatively together on a wide array of policies including education, economics, foreign relations, and domestic and international security. Through a complicated but representative system of democratic voting, the EU establishes rules that govern the member states' rights to free trade along with their citizens' borderless opportunities to travel, work, live, and retire. Nineteen of the countries share a common currency, the euro, and citizens of any one country are also considered citizens of the larger EU.[43]

In recent years, artificial intelligence has become an increased focus of EU policy considerations and legislation. In April 2018, a "Declaration of Cooperation on Artificial Intelligence" was signed by 24 of the then 28 independent members. Within three months, the other four signed on as well. The declaration is a direct and concise statement of the EU member states' agreement to work together on what they consider the most important issues of AI—from maintaining Europe's competitiveness in research, development, and deployment of AI to addressing the social, economic, moral, and legal questions the technology unavoidably creates. Generally, the declaration confirms the member nations' agreement to cooperate on boosting Europe's technology and industrial capacity in AI, addressing the socioeconomic challenges AI will cause, and ensuring that appropriate legal and ethical frameworks are established to protect fundamental human rights—including the privacy and protection of personal data. In particular, the nations agreed to exchange views between them on research strategies and agendas to create a synergy across Europe, to mutually contribute to making AI beneficial for all governments and companies (both small, medium, and large and whether in the AI sector or not), and to ensure that humans remain at the center of AI development, deployment, and decision-making.[44]

At the same time as the "Declaration of Cooperation," the EU's General Data Protection Regulation (GDPR) protocol went into effect.[45] It established an internationally agreed law between all member nations that imposes heavy fines on companies providing AI and/or data in violation of its information-protection guidelines. The law doesn't apply just to companies headquartered in Europe, but also to any company that processes the personal data of EU residents, regardless of whether the data is processed within an EU territory or abroad. With fines as high as 20 million euros, or 4 percent of the violating company's global revenues (whichever is higher), the law can have significant impact. Even for major companies like Google, Facebook, Baidu, Alibaba, Tencent, and Huawei, the business ramifications for any violations of the GDPR could be great, with fines potentially in the billions.[46]

After the "Declaration of Cooperation" and the GDPR became effective, the EU published a communication entitled "Artificial Intelligence for Europe."[47] It summarized and explained the AI perspectives of the member states and their intention to develop specific and detailed initiatives going forward. In the document's introduction, the importance and purposes of the EU's focus were stated, in part, as follows:

> Artificial intelligence (AI) is already part of our lives—it is not science fiction. From using a virtual personal assistant to organise our working day, to travelling in a self-driving vehicle, to our phones suggesting songs or restaurants that we might like, AI is a reality. Beyond making our lives easier, AI is helping us to solve some of the world's biggest challenges: from treating chronic diseases or reducing fatality rates in traffic accidents to fighting climate change or anticipating cybersecurity threats.
>
> . . .
>
> Like the steam engine or electricity in the past, AI is transforming our world, our society and our industry. Growth in computing power, availability of data and progress in algorithms have turned AI into one of the most strategic technologies of the 21st century. The stakes could not be higher. The way we approach AI will define the world we live in. Amid fierce global competition, a solid European framework is needed.[48]

To support its vision for the ethical use of AI, the EU established a high-level group of independent experts on AI to draft guidelines and policy/investment recommendations for all member countries to follow. In April 2019, the group presented its final report entitled "Ethics Guidelines for Trustworthy AI." The document established various criteria that AI systems should meet to remain trustworthy and consistent with Western perspectives of human rights, and it recommended specific methods of assessing AI implementations consistent with those criteria. The entirety of the report was based on the prefatory assessment that AI must: (1) be lawful, ensuring compliance with all applicable laws and regulations; (2) be ethical, ensuring adherence to ethical principles and values; and (3) be sufficiently robust, from technical and social perspectives, to ensure that, even with good intentions, AI systems do not inadvertently cause unintentional harm.[49]

At the same time the EU was beginning to formalize its plans for AI, individual European countries were, on their own, doing the same — most notably the UK and France.

THE UNITED KINGDOM

In various earlier parts of this book, we've discussed the enormous contributions that Great Britain and its institutions, scientists, and citizens have historically made to the development of computing technologies and, more recently, to AI and the specific advance of machine learning algorithms and applications. As with other Western nations currently playing a significant role in the development of AI, the UK is rightly perceived to be primarily concerned with ensuring its ethical use to avoid future compromise and dilemmas. To its credit, the British government's actions and initiatives in recent years have all been consistent with that perception.

In mid-2017, the British Parliament established a select committee on AI to broadly consider the ethical and socioeconomic implications of AI. After a year's work, the committee published a sweeping 183-page report in April 2018 entitled "AI in the UK: Ready, Willing, and Able?" The report acknowledges that the UK will not be able to rival either the sums of investments made by the US or China, the number of individuals with AI talent and expertise available to those two countries, or the practical ability to turn research advantages into commercial realities as readily as they can.[50] But the UK has enjoyed a long and unique record of world-leading AI research capabilities and, again, is a recognized leader on ethical considerations of the technology — with world-renowned institutions like the Alan Turing Institute (the UK's national institute for data science and artificial intelligence),[51] and the acclaimed Future of Humanity Institute at the University of Oxford.[52] In consideration of its niche of ethical expertise, the "AI in the UK" report recommended that the government continue its national develop-ment of academic and industrial AI efforts, increase its national invest-ments, and enhance its educational system's focus on computer and AI fields, not only on technical learning and teaching skills, but also on eth-ical standards. The report also recommended that the British govern-ment itself take an aggressive lead in the development and deployment of AI. As a starting point in the process, and in conclusion to the report, the committee suggested that a national AI code be established, with five overarching principles: (1) artificial intelligence should be developed for the common good and benefit of humanity; (2) artificial intelligence should operate on principles of intelligibility and fairness; (3) artificial intelligence should not be used to diminish the data rights or privacy of individuals, families, or communities; (4) all citizens should have the right to be educated to enable them to flourish mentally, emotionally, and economically alongside artificial intelligence; and (5) the autono-mous power to hurt, destroy, or deceive human beings should never be vested in artificial intelligence.[53]

Former UK prime minister Theresa May summarized her country's intentions and perspectives on AI at the 2018 World Economic Forum in Davos, where she stated, in part:

> [W]e are establishing an Institute of Coding—a consortium of more than 60 universities, businesses and industry experts to support training and retraining in digital skills.
>
>   . . .
>
> And we are establishing the UK as a world leader in Arti-ficial Intelligence, building on the success of British companies like DeepMind.
>
>   . . .
>
> We are absolutely determined to make our country the place to come and set up to seize the opportunities of Artificial Intel-ligence for the future. But as we seize these opportunities of

technology, so we also have to shape this change to ensure it works for everyone — be that in people's jobs or their daily lives.

. . .

But when technology platforms work across geographical boundaries, no one country and no one government alone can deliver the international norms, rules and standards for a global digital world.

Technology companies themselves, investors, and all our international partners need to play their part.

. . .

In a global digital age we need the norms and rules we establish to be shared by all.

. . .

This includes establishing the rules and standards that can make the most of Artificial Intelligence in a responsible way, such as by ensuring that algorithms don't perpetuate the human biases of their developers.[54]

FRANCE

At the Paris AI for Humanity Summit in March 2018, French president Emmanuel Macron announced his country's ambitious new five-year plan to transform France into a global leader in AI research, training, and enterprise. The plan consists of four parts. First, it includes several initiatives to strengthen France's AI ecosystem and attract international AI talent through various means, including the creation of new research institutes across the country. Second, France will fund and support the development of public and private data exchange platforms to drive the application of AI in sectors where it's already a sound leader, like healthcare. Third, the government will create a regulatory and financial framework to develop domestic AI "champions" and will create a European equivalent of America's DARPA. And, fourth, France will invest in research for the development of AI ethics to ensure that it is only deployed in transparent, explainable, and nondiscriminatory ways. In total, the French will invest 1.5 billion euros in AI by the end of their current five-year plan.[55]

CANADA

The Canadian government's support of AI research goes all the way back to 1982, when the Canadian Institute for Advanced Research (CIFAR) was created to promote strategic inquiry in academically complex areas. Among the first areas CIFAR focused on were AI and robotics when it launched the Artificial Intelligence, Robotics, and Society program in 1983.[56]

Despite the AI winters of the late 1980s and early '90s, CIFAR's ongoing support of Canadian universities kept them continuously involved and active in the field. In Chapter 9, we talked about the early focus on machine learning technologies first pursued by Geoffrey Hinton, Yoshua Bengio, and their followers. In 1987, the University of Toronto hired Hinton. In 1993, the University of Montreal hired Bengio and, in 2003, the University of Alberta hired another leading AI researcher (in the field of reinforcement learning), Rich Sutton, an American. The AI ecosystem of those universities and others throughout Canada put the country at the forefront of the artificial intelligence and machine learning revolution, and has kept it there ever since.

In March 2017, Canada became the first nation in the world to announce a formal artificial intelligence plan. Through his "Pan-Canadian Artificial Intelligence Strategy,"[57] Prime Minister Justin Trudeau committed Canada to maintaining its academic leadership with a $125 million investment over five years. The investment funded the creation of three new AI institutes and is primarily directed toward attracting and maintaining researchers and graduates in an effort to continue Canada's long history of leadership related to the research of technical advancements and the study of economic, ethical, and legal issues surrounding AI.[58]

A year later, Canada hosted the 2018 G7 Summit in Quebec,[59] before which Trudeau and French president Macron jointly announced their intent to create an international study group on ethical AI. Later in the year, Canada also held a G7 multi-stakeholder conference on AI in Montreal, which was focused on responsible business adoptions and applications of AI.[60]

AUSTRALIA AND NEW ZEALAND

Australia does not yet have a formal artificial intelligence strategy, but it announced significant funding to support the development of AI in its 2018–2019 national budget.[61] It is currently in the process of creating a national "Technology Roadmap," a "Standards Framework," and a national "AI Ethics Framework" to ensure the responsible development of AI. National budget funds are also allocated to enhance advanced education and other initiatives, especially to increase its supply of national and foreign AI talent.[62]

While developing its formal AI strategy, Australia has taken a very active step to protect its citizens, businesses, and government from outside digital threats. In August 2018, it banned the Chinese telecom service and equipment giants Huawei and ZTE from participating in its 5G infrastructure and from selling 5G equipment on the continent.[63] Malcolm Turnbull, Australia's prime minister at the time the ban was enacted, explained that it was necessary to protect Australia against future risks of Chinese espionage. He pointed out that even though no one in Australia suspected Huawei or ZTE of acting illegally in the context of their relations *at the time*, it isn't possible to fully predict the future. As he said, "Capability takes a long time to put in place. Intent can change in a heartbeat, so, you have to hedge and take into account the risk that can change in the years ahead."[64]

Within three months, the neighboring island of New Zealand followed Australia's lead and likewise blocked Huawei and ZTE 5G technology, attempting to ensure that the two countries' remote location in the southwestern Pacific Ocean will, to the extent currently possible, remain protected from Chinese digital intrusion.[65]

———

Of the 195 countries in the world, fewer than 30 have specific strategies or initiatives in place for AI, and there are less than ten international agreements existing between two or more of them.[66] Fundamentally, when it comes to the new technologies of AI and machine learning applications, there are two broad categories of nations and national agendas: those that approach the development and deployment of AI in ways consistent with Western, democratic ideals of human rights and individual dignities . . . and those that don't. In many ways, with respect to AI, the challenge for democracies is no different than it has always been with regard to any new development of industry or technology of the past. Separate and together, democracies must work to ensure that AI is developed and implemented only in ways that ensure the rights and freedoms to which their citizens are fundamentally entitled . . . and that protect their citizens from all contrary uses, whether domestic or foreign.

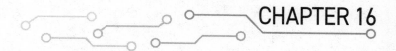

CHAPTER 16

A COMPUTER'S CONCLUSION

Our job is now to convince the public in particular that using AI to achieve these aims is a necessary and desirable part of our society, but we cannot afford to do so unless we know how it will best be used and when. But in the end, the future demands we make moral decisions as we begin to build a world that is truly safe and sustainable, one where humans and AI can truly coexist together.

—GPT-2 (1558 Model)
An OpenAI Language-Generating Neural Network

GPT-2 is a large-scale, unsupervised machine learning application created by the American nonprofit organization, OpenAI.[1] An acronym for *Generative Pre-Training, Version 2*, the GPT-2 application was trained on a data set of eight million web pages and designed as an intricately deep neural network capable of weighing 1.5 billion parameters. Its narrow task is to generate humanlike, written language responses to submissions of text, or "prompts," that it generates in the form of either a proposed *continuation* of the prompt, or a *response* if the prompt was submitted in the form of a question. In essence, GPT-2's function is to create additional words that are: (1) consistent with the patterns and content of new text submissions, and (2) based on patterns the program has discerned from its immense training set of internet information. Consistent with everything we described in Chapter 9 about the machine learning process, GPT-2 performs its task by assessing text submitted to it in the context of all it has learned from its previous experiences—and then it "predicts" the next words that should logically follow.

The program has proven incredibly adept at creating humanlike words, sentences, paragraphs, and even stories. To an impressive extent, it routinely passes the Turing Test—the imitation game that we first spoke of in Chapter 4—by impersonating the language of humans so well that its words are often indistinguishable from ours. As such, the technology has great potential to enable better writing assistants, more capable chat agents, better translation applications, and more efficient and accurate speech-recognition systems. Equally, though, in the wrong hands it could also be used to generate disinformation and biased or abusive language—such as fake news stories, false communications, or impersonated social media posts.

To OpenAI's credit, the organization doesn't develop commercial products or seek profit, and their sole organizational mission is to conduct research and development to ensure AI remains safe and that it "benefits all of humanity."[2] Consistent with their mission, OpenAI generally releases all of their research, their code, and their developments to other research organizations and to the public—hence, the name *Open*AI.

But, with respect to GPT-2, they faced a dilemma, which to this point they've managed quite admirably. Not wanting to make a potentially hazardous AI program available to the public, they chose instead to release GPT-2 in stages, weighing the safety concerns against the public's ability to assess and use the algorithm along the way. They also chose not to release the data set, training code, or model weights by which the algorithm was tuned. Following those parameters, OpenAI first released a staged version of GPT-2, in May 2019, that measured 345 million parameters (the 345M version), instead of 1.5 billion. In August 2019, they released a 774M version that measured, accordingly, 774 million parameters—still only half of the full-scale version and significantly less susceptible to misuse than the full version would be. Then, having apparently seen no strong evidence of misuse, OpenAI released the full-sized, 1.5 billion parameter model (1558M version) in November 2019.[3]

Before closing this book, I thought it would be an appropriate experiment to informally test GPT-2 myself. On the very first page of this book, in the short Author's Note just before the Prologue, I wrote:

Our focus now must be to openly address the current realities of AI to ensure, as well as we can, that it is implemented only in ways consistent with fundamental human dignities . . . and only for purposes consistent with democratic ideals, liberties, and laws.

At this point in your reading, I trust you know how convinced I am of those words. They seemed a perfect choice to test GPT-2, an appropriate sample to see what kind of "continuation" the program would produce.

When I typed and submitted them into the program, its generator displayed a response almost immediately. The words the algorithm created, on its own and in less time than it took to lift my fingers from the keyboard, are shown as the epigraph at the start of this chapter. They're so cogent to the entirety of this book that they bear repeating. So, here they are. This is from an algorithm familiar with eight million web pages, but prompted only by my 43 words:

> Our job is now to convince the public in particular that using AI to achieve these aims is a necessary and desirable part of our society, but we cannot afford to do so unless we know how it will best be used and when. But in the end, the future demands we make moral decisions as we begin to build a world that is truly safe and sustainable, one where humans and AI can truly coexist together.

Impressive? I think so. And I couldn't have said it better myself. In fact, in an instant and in those two sentences, an artificially intelligent program captured the essence of what I've endeavored to make clear through the previous 15 chapters.

———

So, as you read these last paragraphs of *T-Minus AI*, what does it all mean? That's a monumental question, but these are monumental times. Surely, it means more than any one of us alone can say. The full implications and effects of AI and machine learning technologies remain to be seen in the months, years, and even decades ahead.

What I do know, however, is that we're now at an inflection point in the history of the human race. What we do with respect to AI will impact our present, our future, and perhaps our eventual destiny. The strengths of free nations and democratically represented people are, and will always be, their ability to work cooperatively together in order to preserve their individual liberties and ways of life. This is no time to distance ourselves, to be passive or distracted.

In many ways, it's difficult to write a book before a story is complete. And the AI story is of course only beginning. But understanding current AI and its potential, both good and bad, is essential. And caring is crucial. In the days immediately ahead, decisions will be made, or not, that will follow us forever. One thing I've learned in my years with the United States Air Force, particularly while at the Pentagon, is that our best leadership takes its cues from the needs of its ranks. Likewise, democracy takes its direction from the voice of its people. No system is perfect, but democracy gives standing to its people for the fundamental purpose of ensuring their needs are heard and their rights protected. Ideally, those voices are informed and at least generally aware of the conflicts and potential compromises they face. Effective democracy depends upon that.

If this book contributes in any way to a better understanding of AI and an enhanced appreciation of its significance, then I'll have accomplished my mission. It's time for another awakening, a public awareness, and a conscientious consensus. Those who one day look back upon these times should not be left wishing our eyes had been more open.

ACKNOWLEDGMENTS

Appreciation is a wonderful thing. It makes what is excellent in others belong to us as well.

—Voltaire, 1694–1778
French Philosopher and Author

A universal wave of possibilities underlies all we are and all we do. Our reality only emerges from the energy and involvement of the people with whom we entangle. To be most thoughtful and sincere, expressions of gratitude require some clarity of context.

Not long ago, the probability that this book would become a reality was relatively small. In truth, it was only recently that I even landed upon the concept. Once imagined, though, the idea acquired an extraordinary velocity—but only because of exceedingly good fortune amassed along the way, namely the confidence and contributions of exceptional people who chose to believe in the concept and pave a path to its completion.

The process of creating and publishing a nonfiction book has many steps and stages. If all goes well, the timeline flows from an initial concept to a carefully constructed agent query, then to an agency agreement, then to a formal proposal for potential publishers that, ideally, leads to a publishing contract, then back to actually drafting the manuscript, which, once completed, is hopefully submitted by deadline. The manuscript then moves to editorial development, then to copyediting, cover design, interior layout, and typesetting—all of which results in galley proofs, some level of agreed marketing and sales strategies, and, finally, publication. A few of those steps overlap, but most are entirely independent. None are less or more important than another. And, save for the first step or two, all require the backdrop of industry professionals.

In the case of *T-Minus AI*, the first critical connection was with my agent, the renowned Jane Dystel. The daughter of an American publishing icon, Jane has herself earned a storied and rightful status as a luminary figure in the industry. Without her representation and solid advice, nothing here might ever have happened. Most impressive about Jane is her concise, pragmatic approach and strategic clarity: move quickly, but only when best prepared; make decisions expediently when opportunities demand; and, once decisions are made, move unequivocally *onward*. I suspect that describes Jane's approach to most aspects of her life. It's a no-nonsense philosophy, and a battle cry well worth following.

Onward then to my publisher, Glenn Yeffeth of BenBella Books, who was also an extraordinary connection. He was enthusiastic from the start, but most importantly allowed me the freedom to write the book my way. Whereas others might have constrained or altered my structure, thinking it too broad in scope, Glenn took the risk of letting me follow my original vision—trusting that I could sweep from A to Z in some reasoned and lucid manner. Authors often complain, at least to one another, that their publisher took control of their style, their structure, or even their title—yes, publishers have the final say on many more aspects of a finished book than readers have reason to know. But that was never the case with Glenn and his team. Their contributions and insight were immensely constructive and creative, but also, always, cooperative and deferential. My particular thanks go to editor-in-chief Leah Wilson, editor Vy Tran, copy editor Scott Calamar, senior marketing director Jennifer Canzoneri, senior marketing manager Lindsay

Marshall, deputy publisher Adrienne Lang, vendor content manager Alicia Kania, art director Sarah Avinger, deputy production manager Jessika Rieck, and the entire BenBella sales team.

Perhaps BenBella's most significant contribution came when they reached across the Atlantic to bring aboard my developmental editor, Keith Mansfield. Of Oxford University Press and *Superintelligence* book publishing notoriety, Keith is a successful science and technology publisher, editor, and author in his own right. He offered unparalleled levels of constructive insight and suggested narrative elements both subtle and significant. He also became a friend and creative colleague forever. Cheers, Keith, I look forward to our next drink on the Thames.

Those are the people without whom *T-Minus AI* would never have reached bookstore shelves, but others deserve my express appreciation in more elemental ways. Like life itself, writing a book isn't about idly waiting for your inspiration, moment, or message to somehow magically arrive on its own. Nor is it about first efforts being sufficient. It's about hard work, relentlessly writing and rewriting, again and again, banging your head until the words read right, until they reach whatever threshold your individual standards and personal pride demand. Whatever anyone else may think of *T-Minus AI*, in the end it passed my own acid test. Einstein is attributed with saying, "You do not really understand something unless you can explain it to your grandmother." I think that's true. But, had Einstein known my own grandmother, he would have altered his words slightly. A more precise adage would be "Your grandmother is likely the smartest person you'll ever encounter . . . so if she doesn't understand your explanation, it's sure that no one else will either." So, to you, Grandma, I hope you enjoy the read. You've always been my inspiration and my benchmark. In your ninety plus years of life, you've seen and learned things I still can't imagine. Our late-night conversations have meant more than you know, and my path, my career, this book—none of them would have happened without your influence. Thank you!

To my sister, Sarah, you are my touchstone. Don't ever let distance cause doubt. Whatever I do, I think in some way of you. Thank you for your strength during all of those days behind us. The struggles shared were the struggles we overcame. Time and distance may have grown between us, but the beats of our hearts are near as ever . . . and our common essence compels us to tighten our ties in all of our tomorrows.

To you, Dylan, my not-so-little-anymore brother, our lives are forever entangled. We are iterations of the same soul, merely split into separate but identical parts. The wave that pushes us forever forward can never collapse, and the underlying state will keep us always united. We're ubiquitous, nothing without each other, both here and there, and your perspective will always matter.

To Philip, a better cousin couldn't be had. Thanks for your time, conversation, and confirmations.

A special thanks to Lt. General "Dash" Jamieson, without whom my path would surely have taken some other direction, leading to who knows where. My years with you were everything, but as our pages turn, exciting new chapters still remain for each of us to write. The world is yours. You've earned it.

To Elsa Kania and Lindsey Sheppard, your knowledge and gracious input were crucial to seminal sections of the book. Gracias. So many people and circumstances are incredibly better because of the expertise and altruism you both so often share.

To Svetlana, your extraordinary sense of style, design, and self is unmatched. Thank you for that first AI "101" brief. Hindsight being 20/20, it was the catalyst for so much.

Likewise, thank you to all who have significantly shaped my thinking, my career, and my opportunities—and who've also suffered my idiosyncrasies as costs along the way. Among others, they include General Stephen "Seve" Wilson, Lt. General John N. T. "Jack" Shanahan, Colonel Jason "Jobu" Brown, Colonel Drew Cukor, Josh Marcuse, Aaron Schumacher, Bryan Lewis, Joe Larson, Dr. Heather Roff, Kara Frederick, Lloyd Dabbs, Brendan McCord, Group Captain Blythe Crawford, Joy Shanaberger, Kristina McTigue, Lt. Col. Tommy Jones, Ryan "Shaq" O'Neal, Sheena Puleali'i, Tyler Johnson, and Liz Vaughan Moyer.

To Aaron, and my second family of Pam, Joe, Al, and the entire Ennest/Dobis clan—thanks for your friendship, support, encouragement, and loyalty over all the years. You complete my roots and strengthen whatever is possible. I hope I contribute in some way to you and your lives, for you surely augment me and mine.

Last, to my best friend. Dad, none of this would ever have been possible without you. You told me long ago that this would all be worth it. You're right, it is. And it is you I appreciate most.

NOTES

1. In a 45-minute open lesson to mark the start of the Russian school year,
   Putin addressed various topics including AI, space, medicine, and the
   capabilities of the human brain. The session was broadcast throughout the
   country to an estimated audience of over one million. James Vincent, "Putin
   Says the Nation That Leads in AI 'Will Be the Ruler of the World,'" The
   Verge, September 4, 2017, www.theverge.com/2017/9/4/16251226/russia
   -ai-putin-rule-the-world. See also: Staff, "Whoever Leads in AI Will Rule
   the World: Putin to Russian Children on Knowledge Day," RT Question
   More, September 1, 2017, www.rt.com/news/401731-ai-rule-world-putin/.
2. Mike Wall, "Huge Asteroid 'Florence' Zooms Past Earth in Record-
   Breaking Flyby," Space.com, September 1, 2017, www.space.com/38013
   -huge-asteroid-florence-makes-earth-flyby.html.
3. The Chinese Next Generation Artificial Intelligence Plan will be discussed
   in various parts of this book, particularly Chapters 12 and 13. For a full
   translation of the plan, see: Translated by Graham Webster, Paul Triolo,
   Elsa Kania, and Rogier Creemers, "A Next Generation Artificial Intelligence Development Plan," China Copyright and Media, July 20, 2017,
   https://chinacopyrightandmedia.wordpress.com/2017/07/20/a-next
   -generation-artificial-intelligence-development-plan/.

CHAPTER 2

1. Richard B. Larson and Volker Bromm, "The First Stars in the Universe," *Scientific American*, January 19, 2009, www.scientificamerican.com/article/the-first-stars-in-the-un.
2. See: Michael Marshall, "The Event That Transformed Earth, BBC, July 2, 2015, www.bbc.com/earth/story/20150701-the-origin-of-the-air-we-breathe.
3. Paleontologists now believe a 437-million-year-old fossil of a prehistoric scorpion is the oldest known example of an air-breathing animal with the capability of crawling onto land. At the time this species lived, the only other living things on land would have been plants and proto-plants. The discovery dates the 2-centimeter-long scorpion to be 16–17 million years older than the oldest known millipede fossil, which was previously believed to have been the oldest fossilized evidence of a land animal. All of this would have been 10–20 million years before the first fish emerged from ancient seas. Laura Arenschield, "Fossil Is the Oldest-Known Scorpion," *Ohio State News*, January 16, 2020, https://news.osu.edu/fossil-is-the-oldest-known-scorpion/. See also: Katie Hunt, "Prehistoric Scorpion Is Earliest Known Animal to Venture from Sea to Land," CNN, January 16, 2020, https://www.cnn.com/2020/01/16/world/prehistoric-scorpion-first-to-breathe-on-land-scn/index.html.
4. Mary Bagley, "Devonian Period: Climate, Animals & Plants," Live Science, February 22, 2014, www.livescience.com/43596-devonian-period.html.
5. For decades, there has been an active debate by paleontologists over dinosaur biology and thermal regulation. It was originally assumed that dinosaurs were ectothermic, meaning they were cold-blooded, and regulated their body heat by external sources like the sun and air temperature. But, as skeletal research started to prove that dinosaurs were capable of fast and nimble movement, it became more commonly argued that they were mesothermic, capable of regulating their body temperatures by both internal systems and external forces. A third school contends that evidence supports the argument that dinosaurs were actually endothermic, like mammals, and regulated their blood temperature solely by internal mechanisms. Jen Viegas, "Dinosaurs May Have Been Warm-Blooded," Seeker, May 28, 2015, www.seeker.com/dinosaurs-may-have-been-warm-blooded-1769885183.html.
6. Analysis of recent geologic evidence from the meteor's impact crater—made available only recently from commercial oil drilling data—has now

enabled researchers to quantify, rather than simply speculate about, the energy produced by the meteor's impact. The massive redistribution of geological material and sediment indicates the actual energy that the meteor generated, equaling 100 teratons of TNT. Jane Palmer, "We Finally Know How Much the Dino-Killing Asteroid Reshaped Earth," *Smithsonian.com*, February 5, 2016, www.smithsonianmag.com/science-nature/we-finally -know-how-much-dino-killing-asteroid-reshaped-earth-180958222/. See also: Robert Lee Hotz, "Scientists Discover New Evidence of the Asteroid That Killed Off the Dinosaurs," *Wall Street Journal*, September 9, 2019, www.wsj.com/articles/scientists-discover-new-evidence-of-the-asteroid -that-killed-off-the-dinosaurs-11568055601. See also: *Nova,* "Day the Dinosaurs Died," directed by Sarah Holt, PBS, 2017.

7. It is now commonly accepted that all species of birds descend from dinosaurs. Emily Singer, "How Dinosaurs Shrank and Became Birds," *Scientific American* (from *Quanta Magazine*), June 12, 2015, www.scientificamerican .com/article/how-dinosaurs-shrank-and-became-birds/.

8. Nick Pyenson, *Spying on Whales* (New York: Viking Penguin, 2018).

9. Ann Gibbons, "The Human Family's Earliest Ancestors," *Smithonian. com*, March 2010, www.smithsonianmag.com/science-nature/the-human -familys-earliest-ancestors-7372974/.

10. *National Geographic,* "Why Am I a Neanderthal?," Genographic Project, accessed October 28, 2019, https://genographic.nationalgeographic.com /neanderthal/. See also: Emma Groeneveld, "Denisovan," Ancient History Encyclopedia, May 5, 2019, www.ancient.eu/Denisovan/.

11. Ian Tattersol, "Homo sapiens," *Encyclopedia Britannica*, accessed May 20, 2019, www.britannica.com/topic/Homo-sapiens.

12. "How Did Language Begin?," Linguistic Society of America, accessed May 20, 2019, www.linguisticsociety.org/resource/faq-how-did-language-begin.

13. Joshua J. Mark, "Writing," Ancient History Encyclopedia, April 28, 2011, www.ancient.eu/writing/.

14. Laura Shumm, "Who Created the First Alphabet?," History, updated August 18, 2018, www.history.com/news/who-created-the-first-alphabet.

15. Steven Moran and Daniel McCloy, eds., Phoible 2.0, Jena: Max Planck Institute for the Science of Human History, 2019, http://phoible.org.

16. Christopher McFadden, "The Invention and History of the Printing Press," Interesting Engineering, September 12, 2018, https://interesting engineering.com/the-invention-and-history-of-the-printing-press.

CHAPTER 3

1. Summit is a supercomputer developed by IBM for use at Oak Ridge National Laboratory (Tennessee), which, as of August 2019, was the fastest supercomputer in the world, having achieved that rank in November 2018. Summit is capable of 200 petaflops, or 200 quadrillion calculations per second. Its racks of equipment occupy a space the size of two tennis courts, are connected by over 185 miles of fiber-optic cable, and circulate 4,000 gallons of water a minute to cool the machine's 37,000 processors. "The Most Powerful Computers on the Planet," IBM Summit, accessed August 23, 2019, www.ibm.com/thought-leadership/summit-supercomputer/.

2. Dr. Y., "The Ishango Bone, Cradle of Ancient Mathematics," *African Heritage*, August 29, 2013, https://afrolegends.com/2013/08/29/the-ishango -bone-craddle-of-mathematics/.

3. "Babylonian and Mesopotamian Mathematics," Facts and Details, accessed August 23, 2019, http://factsanddetails.com/world/cat56/sub402/entry -6083.html.

4. "Easy as 1, 2, 3 . . . ," The Renaissance Mathematicus, May 3, 2018, https:// thonyc.wordpress.com/2018/05/03/as-easy-as-123/.

5. For a deeper description of scientific notation, see: Janet Shiver, PhD, and Teri Willard, EdD, "Scientific Notation and Order of Magnitude," Visionlearning, accessed August 29, 2019, www.visionlearning.com/en/library /Math-in-Science/62/Scientific-Notation-and-Order-of-Magnitude/250.

CHAPTER 4

1. Marissa Schwartz, "Delimitations: Rethinking the U.S.-Mexican Border," *Time*, November 20, 2014, http://time.com/3813321/us-mexico-border -1821/.

2. History.com editors, "Texas Declares Independence," History, July 21, 2010, www.history.com/this-day-in-history/texas-declares-independence.

3. History.com editors, "Texas Declares Independence."

4. Treaty of Peace, Friendship, Limits and Settlement between the United States of America and the Mexican Republic (unofficially, The Treaty of Guadalupe Hidalgo), Articles XII–XV, February 2, 1848, www.archives .gov/education/lessons/guadalupe-hidalgo. See also: Christopher Minster, "The Treaty of Guadalupe Hidalgo," ThoughtCo., May 30, 2019, www .thoughtco.com/the-treaty-of-guadalupe-hidalgo-2136197.

5. Minster, "The Treaty of Guadalupe Hidalgo." See also: NCC staff, "On the Day, the Treaty of Guadalupe Hidalgo Is Signed," National Constitution Center, February 2, 2019, https://constitutioncenter.org/blog/on-the -day-the-treaty-of-guadalupe-hidalgo-is-signed.

6. "The Zimmermann Telegram," National Archives, accessed August 23, 2019, www.archives.gov/education/lessons/zimmermann.

7. Gordon Corera, "Why Was the Zimmermann Telegram So Important?," BBC News, January 17, 2017, www.bbc.com/news/uk-38581861.

8. Corera, "Why Was the Zimmermann Telegram So Important?"

9. History.com editors, "Zimmermann Telegram Published in the United States," History, November 5, 2009, www.history.com/this-day-in-history /zimmermann-telegram-published-in-united-states.

10. "What Was the Enigma Machine?," World Atlas, accessed August 23, 2019, www.worldatlas.com/articles/what-was-the-enigma-machine.html.

11. See: "The History of Enigma," Crypto Museum, accessed August 20, 2019, www.cryptomuseum.com/crypto/enigma/hist.htm.

12. Karleigh Moore, Ethan W., and Ejun Dean, "Enigma Machine," *Brilliant*, accessed May 30, 2019, https://brilliant.org/wiki/enigma-machine/.

13. "The History of Enigma," Crypto Museum.

14. Karla Adam, "Alan Turing, a Founding Father of Computer Science, Revealed as New Face of British 50-Pound Note," *Washington Post*, July 15, 2019, www.washingtonpost.com/world/europe/alan-turing-a-founding -father-of-computer-science-revealed-as-new-face-of-british50-pound -note/2019/07/15/96a1e46a-a6ff-11e9-86dd-d7f0e60391e9_story.html?utm _term=.24056a5d61e0.

15. Chris Smith, "Cracking the Enigma Code: How Turing's Bombe Turned the Tide of WWII," BT, November 2, 2017, http://home.bt.com/tech -gadgets/cracking-the-enigma-code-how-turings-bombe-turned-the-tide -of-wwii-11363990654704.

16. Frank Carter, "The Turing Bombe," *Rutherford Journal*, accessed May 31, 2019, www.rutherfordjournal.org/article030108.html. See also: "Bombe," Crypto Museum, accessed August 23, 2019, www.cryptomuseum.com /crypto/bombe/.

17. Although Enigma was used by the Germans for all chain of command and operational radio transmissions throughout the end of the war, shortly before its conclusion Hitler and his most senior commanders began using a new and even more challenging encryption machine for only the highest-level communications among themselves. Called Lorenz, the encoding process it

utilized was far more complex than Enigma, and its design, once again, created exponentially more output variations than were possible before. While Turing remained focused on Enigma, another mathematician at Bletchley Park named Tommy Flowers was assigned to solve the Lorenz problem. To do so, Flowers relied upon a prewar, 1936 paper written by Turing in which he hypothesized and described a computer concept that later came to be called the "Turing Machine." Flowers constructed an interpretation of Turing's imagined machine that essentially did the same thing as the Bombe, but much more robustly. The new machine worked with vacuum tubes and was capable of performing counting operations in roughly the same way as the transistors used in today's digital computers—and that we discuss in Chapter 5. Flowers called his new machine Colossus, and it is appropriately now considered the world's first electrically programmable digital computer. For almost four decades, however, the existence of Colossus and of Flowers's accomplishment in building it remained classified and completely secret, known only to the highest ranks of British intelligence personnel. In light of the British military's secrecy over Colossus, an American machine built between 1943 and 1945 at the University of Pennsylvania was instead given credit at the time as being the first programmable digital computer. Named ENIAC (Electronic Numerical Integrator and Computer), it had been designed and used primarily to calculate artillery firing tables and solutions for the US Army's Ballistic Research Laboratory. ENIAC was given extensive coverage by the press, who captured the world's attention by heralding ENIAC as a "Giant Brain." ENIAC had a speed close to 1,000 times faster than the electromechanical Bombe machines of Bletchley Park. It was also programmable, which meant that it could execute many thousands more calculations per second than any other machine before it, except for the still-secret Colossus. ENIAC could calculate an artillery trajectory in 1/2400th the time it took a human to make the same calculation. Accordingly, one hour of ENIAC operation equated to 2,400 man-hours, or 240 ten-hour workdays. By comparison to today's most basic computers, that was still infinitesimally slow, but by mid-1940s standards, it was unimaginably fast. ENIAC was eight feet tall, almost 100 feet long, and weighed more than 30 tons. Today, a single microchip, much smaller than a fingernail, has far more computing capacity than those 30 tons of hardware.

18. Jack Copeland, "Alan Turing: The Codebreaker Who Saved 'Millions of Lives,'" BBC News, June 19, 2012, www.bbc.com/news/technology -18419691.

19. A. M. Turing, "Computing Machinery and Intelligence," 1950, www.csee .umbc.edu/courses/471/papers/turing.pdf

20. Turing, "Computing Machinery and Intelligence."

21. Turing, "Computing Machinery and Intelligence."

22. In 1947, Bell Laboratories introduced the first transistor. Priya Ganapati, "Dec. 23, 1947: Transistor Opens Door to Digital Future," *Wired*, December 23, 2009, www.wired.com/2009/12/1223shockley-bardeen-brattain -transistor/. In 1958, the first integrated circuit was developed, allowing large numbers of small transistors to be placed on a single, small "chip" of semi-conductor material. "The History of the Integrated Circuit," AnySilicon, March 27, 2017, https://anysilicon.com/history-integrated-circuit/.

CHAPTER 5

1. *Ethnologue*, a widely used academic and commercial source on world languages, places the 2019 number of spoken languages at 7,111, with 3,995 corresponding written languages. David Eberhard, Gary Simons, and Charles Fennig, eds., *Ethnologue: Languages of the World,* 22nd ed., "Frequently Asked Questions," accessed August 16, 2019, www.ethnologue .com/faq.

2. Eberhard et al., *Ethnologue*, "What Are the Top 200 Most Spoken Languages?," accessed August 16, 2019, www.ethnologue.com/statistics/size.

3. Eberhard et al., "What Are the Top 200 Most Spoken Languages?"

4. Google Translate, accessed August 23, 2019, https://translate.google.com /intl/en/about/languages/.

5. "History of FORTRAN Language," LivePhysics, accessed August 20, 2019, www.livephysics.com/computational-physics/fortran/history-fortran -language/.

6. See: Andy Alfred, "May 1, 1964: First Basic Program Runs," *Wired*, May 1, 2008, www.wired.com/2008/05/dayintech-0501-2/.

7. Manisha Priyadarshini, "10 Most Popular Programming Languages in 2019: Learn to Code," Fossbytes, August 24, 2019, https://fossbytes.com /most-popular-programming-languages/. See also: "Top 7 Most Popular Programming Languages (Most Used High Level List)," Software Testing Help, August 21, 2019, www.softwaretestinghelp.com/top-programming -languages/.

8. ASCII Code—the Extended ASCII Table, accessed August 23, 2019, www .ascii-code.com/.

9. Stephanie Pappas, "Facts About Silicon," Live Science, April 27, 2018, www.livescience.com/28893-silicon.html.

10. Michael Feldman, "Intel Gives Moore's Law a Makeover," The Next Platform, May 13, 2019, www.nextplatform.com/2019/05/13/intel-gives -moores-law-a-makeover/.

11. "1965: 'Moore's Law' Predicts the Future of Integrated Circuits," Computer History Museum/The Silicon Engine, www.computerhistory.org /siliconengine/moores-law-predicts-the-future-of-integrated-circuits/.

12. Feldman, "Intel Gives Moore's Law a Makeover."

CHAPTER 6

1. NIH Staff, "Brain Basics: Know Your Brain," National Institute of Neurological Disorders and Stroke, accessed August 23, 2019, www.ninds.nih .gov/Disorders/Patient-Caregiver-Education/Know-Your-Brain.

2. See: Mary Bellis, "The History of Mr. Potato Head," ThoughtCo, accessed August 29, 2019, www.thoughtco.com/history-of-mr-potato-head-1992311.

3. Wayne Wu, "The Neuroscience of Consciousness," *The Stanford Encyclopedia of Philosophy* (Winter 2018 ed.) October 9, 2018, https://plato .stanford.edu/entries/consciousness-neuroscience/.

4. See: Yohan John, "Why Science Will Probably Never Address the Problem of Consciousness," *Forbes*—Quora, September 16, 2016, www.forbes .com/sites/quora/2016/09/16/why-science-will-probably-never-address -the-problem-of-consciousness/#198d87e8654b. See also: Michael Shermer, "Will Science Ever Solve the Mysteries of Consciousness, Free Will and God?," *Scientific American*, July 1, 2018, www.scientificamerican .com/article/will-science-ever-solve-the-mysteries-of-consciousness-free -will-and-god/. And see: Kaleigh Rogers, "Scientists Still Don't Totally Understand What It Means to Be Conscious," Vice, September 6, 2018, www.vice.com/en_us/article/paw7nv/what-is-consciousness.

5. Michio Kaku, *The Future of the Mind* (New York: Doubleday, 2014).

6. Kaku, *The Future of the Mind.*

7. Kaku, *The Future of the Mind.*

8. Although some contend the human brain has 100 billion neurons, 86 billion has recently been shown to be a more accurate, average number. Kendra Cherry, "How Many Neurons Are in the Brain?," Verywell Mind, June 11, 2019, www.verywellmind.com/how-many-neurons-are-in-the -brain-2794889. See also: Tanya Lewis, "Human Brain: Facts, Function &

Anatomy," Live Science, September 28, 2018, www.livescience.com/29365
-human-brain.html.

9. Jon H. Kaas, "Neocortex in Early Mammals and Its Subsequent Variations," US National Library of Medicine–National Institutes of Health,
November 23, 2016, www.ncbi.nlm.nih.gov/pmc/articles/PMC3840914/.

10. Stephen L. Chew, "Myth: We Only Use 10% of Our Brain," Association
for Psychological Science, accessed May 31, 2019, www.psychological
science.org/uncategorized/myth-we-only-use-10-of-our-brains.html.

11. Tia Ghose, "The Human Brain's Memory Could Store the Entire Internet," Live Science, February 18, 2016, www.livescience.com/53751-brain
-could-store-internet.html.

12. Luis Villazon, "How Much Energy Does Thinking Use?," *Science Focus*,
accessed May 31, 2019, www.sciencefocus.com/the-human-body/how-much
-energy-does-thinking-use/.

13. Alexandra Ossola, "The Human Brain Could Store 10 Times More Memories Than Previously Thought," *Popular Science*, June 21, 2016, www.popsci
.com/human-brain-could-store-10-times-more-memories-than-previously
-thought.

14. Telegraph Reporter, "Human Brain Can Store 4.7 Billion Books—Ten Times
More Than Originally Thought," *The Telegraph*, January 21, 3016, www
.telegraph.co.uk/news/science/science-news/12114150/Human-brain-can
-store-4.7-billion-books-ten-times-more-than-originally-thought.html.

CHAPTER 7

1. Laura Wise, "The History of Chess and Origins of Chess in Ancient
India," Bright Hub Education, accessed August 24, 2019, www.brighthub
education.com/history-homework-help/122242-from-ancient-india-to
-the-renaissance-history-of-chess/. See also: "Origins of Chess," Chess
.com, December 11, 2008, www.chess.com/article/view/origins-of-chess.

2. Susan Fourtané, "The Turk: Wolfgang von Kempelen's Fake Automaton
Chess Player," Interesting Engineering, August 31, 2018, https://interesting
engineering.com/the-turk-fake-automaton-chess-player.

3. Fourtané, "The Turk."

4. Edward Scimia, "Chess Immortals: The World Champions of Chess," The
Spruce Crafts, January 12, 2019, www.thesprucecrafts.com/world-chess
-champions-611274. See also: "List of World Chess Championships,"

Wikipedia, accessed August 24, 2019, https://en.wikipedia.org/wiki/List
_of_World_Chess_Championships.

5. "List of World Chess Championships," Wikipedia.

6. Edward Winter, "Alekhine's Death," Chess Notes, 2003, www.chesshistory
.com/winter/extra/alekhine3.html. See also: "Alekhine's Death—An Unre-
solved Mystery," ChessBase, March 25, 2006, https://en.chessbase.com
/post/alekhine-s-death-an-unresolved-mystery-. And see: Larry Evans,
"Alekhine's Death Shrouded in Mystery," *Sun Sentinel*, October 26, 1997,
www.sun-sentinel.com/news/fl-xpm-1997-10-26-9710210193-story.html.

7. "Fischer vs Spassky 1972, the Match of the Century," ChessGames.Com,
accessed August 24, 2019, www.chessgames.com/perl/chess.pl?tid=54397.

8. Robert McFadden, "Fischer Loses Title to Russian, 23, by Default," *New
York Times*, April 4, 1975, www.nytimes.com/1975/04/04/archives/fischer
-loses-chess-title-to-russian-23-by-default-fischer-sought-a.html.

9. Bryan Smith, "Kasparov v Karpov / World Chess Championship
1984," Chess.com, June 12, 2018, www.chess.com/article/view/clash-of
-champions-kasparov-karpov.

10. "Kasparov–Karpov World Championship Match 1985," Chess Essen-
tials, June 15, 2018, https://chessentials.com/kasparov-karpov-world
-championship-match-1985/.

11. Scimia, "Chess Immortals." Also note that Kasparov's world title with
FIDE was stripped in 1993 when he and challenger Nigel Short split from
FIDE, the official world governing body of chess, by playing their title
match under the authority of the new Professional Chess Association.

12. William Harston, "Kasparov Puts Clear Blue Water Between Man and
Machine," *Independent*, May 5, 1997, www.independent.co.uk/news
/world/kasparov-puts-clear-blue-water-between-man-and-machine
-1259902.html.

13. Stan Levy, "Man vs. Machine," *Newsweek*, May 4, 1997, www.newsweek
.com/man-vs-machine-173038.

14. Natalie Wolchover, "FYI: How Many Different Ways Can a Chess Game
Unfold?," *Popular Science*, December 15, 2010, www.popsci.com/science
/article/2010-12/fyi-how-many-different-ways-can-chess-game-unfold/.

15. Harston, "Kasparov Puts Clear Blue Water Between Man and Machine."

16. Levy, "Man vs. Machine."

17. For a photo of the March 5, 1997, *Newsweek* cover, see: http://faculty.ycp
.edu/~dweiss/images/kasparov.jpg.

18. Levy, "Man vs. Machine."

19. "I.B.M.'s Stock Surges by 3.6%," *New York Times*, May 13, 1997, www
.nytimes.com/1997/05/13/business/ibm-s-stock-surges-by-3.6.html.

20. Rachel Deason, "A Brief History of Go: The Oldest Board Game in the
World," Culture Trip, December 18, 2017, https://theculturetrip.com
/asia/china/articles/a-brief-history-of-go-the-oldest-board-game-in-the
-world/.

21. "Number of Possible Go Games," Sensei's Library, accessed August 24,
2019, https://senseis.xmp.net/?NumberOfPossibleGoGames. See also:
"Go and Mathematics," Wikipedia, accessed August 24, 2019, https://
en.wikipedia.org/wiki/Go_and_mathematics. And see: "Number of Go
Games with Exactly *n* Moves," The On-Line Encyclopedia of Integer
Sequences/OEIS, accessed August 24, 2019, http://oeis.org/A048289.

22. Mike Murphy, "Google's AI Just Cracked the Game That Supposedly No
Computer Could Beat," Quartz, January 27, 2016, https://qz.com/603313
/googles-ai-just-cracked-the-game-that-supposedly-no-computer-could
-beat/.

23. Sam Shead, "Google's $500+ Million Purchase of DeepMind Just Got Very
Interesting," *Business Insider*, July 21, 2016, www.businessinsider.com
/googles-400-million-acquisition-of-deepmind-is-looking-good-2016-7.

24. "The Google DeepMind Challenge Match," March 2016, DeepMind,
accessed August 24, 2019, https://deepmind.com/research/alphago
/alphago-korea/.

25. *AlphaGo*, directed by Greg Kohs (US: Moxie Pictures and Reel As Dirt,
2017).

26. Christopher Moyer, "How Google's AlphaGo Beat a Go World Champion," *The Atlantic*, March 28, 2016, www.theatlantic.com/technology
/archive/2016/03/the-invisible-opponent/475611/.

27. *AlphaGo*, directed by Greg Kohs.

28. Jack Crosbie, "The Strange Beauty of AlphaGo," Inverse, April 25, 2017,
www.inverse.com/article/30681-alphago-documentary-tribeca-film
-festival.

29. Moyer, "How Google's AlphaGo Beat a Go World Champion."

30. *AlphaGo*, directed by Greg Kohs.

31. *AlphaGo*, directed by Greg Kohs.

32. Jessica Cussins, "DeepMind's AlphaGo Zero Becomes Go Champion
Without Human Input," Future of Life Institute, October 18, 2017,
https://futureoflife.org/2017/10/18/deepminds-alphago-zero-becomes-go
-champion-without-human-assistance/.

33. David Bloom, "What Olympic Recognition Could Mean for Esports, and Vice Versa," *Forbes*, October 18, 2018, www.forbes.com/sites /dbloom/2018/10/18/esports-olympics-recognition-ioc-esl-advertising -sponsors/#7ca6927d3b05.

34. OpenAI, accessed May 24, 2019, https://openai.com/.

35. *Starcraft II: Wings of Liberty* is a science fiction, strategy-based video game produced by Blizzard Entertainment, an American company that, as of 2018, had approximately 9,900 employees operating in 20 countries around the world. See: Activision Blizzard 2018 SEC Form 10-K Annual Report, accessed August 24, 2019, https://investor.activision.com/node/32301 /html.

36. Sam Rutherford, "5 Things to Know About Tencent, the Chinese Internet Giant That's Worth More Than Facebook Now," Gizmodo, November 27, 2017, https://gizmodo.com/5-things-to-know-about-tencent-the-chinese -internet-gi-1820767339.

CHAPTER 8

1. "The Cosmic Background Microwave," Planck Mission, accessed August 24, 2019, http://planck.cf.ac.uk/science/cmb.

2. Brian Greene, "The Detection of Gravitational Waves Was a Scientific Breakthrough, but What's Next?," *Smithsonian Magazine*, April 2016, www.smithsonianmag.com/science-nature/detection-gravitational-waves -breakthrough-whats-next-180958511/.

3. Ota Lutz, "How Scientists Captured the First Picture of a Black Hole," Jet Propulsion Laboratory, April 19, 2019, www.jpl.nasa.gov/edu/news/2019/4/19 /how-scientists-captured-the-first-image-of-a-black-hole/.

4. Leonard Kleinrock, "Information Flow in Large Communication Nets, Proposal for a Ph.D. Thesis," as submitted to Massachusetts Institute of Technology Committee on Graduate Study and Research, Electrical Engineering Department on May 31, 1961, and approved on July 24, 1961, www.lk.cs.ucla.edu/data/files/Kleinrock/Information%20Flow%20in %20Large%20Communication%20Nets.pdf.

5. Guy Raz, "'Lo' and Behold: A Communication Revolution," National Public Radio, October 29, 2009, www.npr.org/templates/story/story.php ?storyId=114280698.

6. Kim Ann Zimmermann and Jesse Emspak, "Internet History Timeline: ARPANET to the World Wide Web," Live Science, June 27, 2017, www .livescience.com/20727-internet-history.html.

7. Tim Berners-Lee, "Information Management: A Proposal," w3.org, accessed August 24, 2019, www.w3.org/History/1989/proposal.html.

8. Alyson Shontell, "Flashback: This Is What the First-Ever Website Looked Like," *Business Insider*, June 29, 2011, www.businessinsider.com/flashback -this-is-what-the-first-website-ever-looked-like-2011-6.

9. "Internet of Things (IoT) Connected Devices Installed Base World- wide from 2015 to 2025 (in Billions)," Statista, accessed May 25, 2019, www.statista.com/statistics/471264/iot-number-of-connected-devices -worldwide/.

10. Bernard Marr, "How Much Data Do We Create Every Day? The Mind- Blowing Stats Everyone Should Read," *Forbes*, May 21, 2108, www.forbes .com/sites/bernardmarr/2018/05/21/how-much-data-do-we-create-every -day-the-mind-blowing-stats-everyone-should-read/#6e8cf5d760ba.

11. Marr, "How Much Data Do We Create Every Day?"

12. "Total Number of Websites," Internet Live Stats, accessed August 25, 2019, www.internetlivestats.com/total-number-of-websites/.

13. Ingrid Lunden, "Amazon Share of the US e-Commerce Market Is Now 49%, or 5% of All Retail Spend," TechCrunch, July 2018, https://tech crunch.com/2018/07/13/amazons-share-of-the-us-e-commerce-market-is -now-49-or-5-of-all-retail-spend/.

14. "Alibaba Tops e-Commerce Market Share While Facing Fresh Competi- tion in China, Marketing," Marketing, April 2018, https://old.marketing -interactive.com/alibaba-tops-e-commerce-market-share-while-facing -fresh-competition-in-china/.

15. Our World in Data, accessed September 12, 2019, https://ourworldindata .org/.

16. Our World in Data.

17. Kevin Fogarty, "Where Did 'Cloud' Come From?," ITWorld, May 14, 2012, www.itworld.com/article/2726701/where-did--cloud--come-from -.html.

18. Bob Evans, "#1 Microsoft Beats Amazon in 12-Month Cloud Rev- enue, $26 Billion to $23.4 Billion: IMB Third," *Forbes*, October 29, 2018, www.forbes.com/sites/bobevans1/2018/10/29/1-microsoft-beats -amazon-in-12-month-cloud-revenue-26-7-billion-to-23-4-billion-ibm -third/#5e2a208f2bf1.

CHAPTER 9

1. See: University of Queensland, "History of Artificial Intelligence," accessed August 25, 2019, https://qbi.uq.edu.au/brain/intelligent-machines/history-artificial-intelligence.
2. University of Queensland, "History of Artificial Intelligence."
3. Dave Gershgorn, "The Inside Story of How AI Got Good Enough to Dominate Silicon Valley," Quartz, June 18, 2018, https://qz.com/1307091/the-inside-story-of-how-ai-got-good-enough-to-dominate-silicon-valley/.
4. Cade Metz, "Turing Award Won by 3 Pioneers in Artificial Intelligence," *New York Times*, March 27, 2019, www.nytimes.com/2019/03/27/technology/turing-award-ai.html.
5. Nick Bostrom, *Superintelligence: Paths, Dangers, Strategies* (England: Oxford University Press, 2014).

CHAPTER 10

1. "List of Cognitive Biases," Wikipedia, accessed August 24, 2019, https://en.wikipedia.org/wiki/List_of_cognitive_biases.
2. James Risley, "Microsoft's Millennial Chatbot Tay.ai Pulled Offline After Internet Teaches Her Racism," GeekWire, March 24, 2016, www.geekwire.com/2016/even-robot-teens-impressionable-microsofts-tay-ai-pulled-internet-teaches-racism/.
3. Sophie Kleeman, "Here Are the Microsoft Twitter Bot's Craziest Racist Rants," Gizmodo, March 24, 2016, https://gizmodo.com/here-are-the-microsoft-twitter-bot-s-craziest-racist-ra-1766820160.
4. "Microsoft 'Deeply Sorry' for Racist and Sexist Tweets by AI Chatbot," *The Guardian*, March 26, 2016, www.theguardian.com/technology/2016/mar/26/microsoft-deeply-sorry-for-offensive-tweets-by-ai-chatbot.
5. Jeffrey Dastin, "Amazon Scraps Secret AI Recruiting Tool That Showed Bias Against Women," Reuters, October 9, 2018, www.reuters.com/article/us-amazon-com-jobs-automation-insight/amazon-scraps-secret-ai-recruiting-tool-that-showed-bias-against-women-iduskcn1mk08g.
6. See: "Selection Bias," Stat Trek, accessed August 25, 2019, https://stattrek.com/statistics/dictionary.aspx?definition=selection_bias.
7. For instance, see: Dave Gershgorn, "Google Created a Tool to Test for Biases in AI Data," Quartz, September 13, 2018, https://qz.com/1389859/google-created-a-tool-to-test-for-biases-in-ai-data/.

CHAPTER 11

1. For an interesting and concise explanation of simple machines and related information, see: Michelle Nickolaisen, "Examples of Simple Machines & Complex Machines," Sciencing, April 28, 2018, https://sciencing.com /examples-machines-amp-complex-machines-7221376.html.
2. Martin Armstrong, "Rise of the Industrial Robots," Statista, October 9, 2017, www.statista.com/chart/11397/rise-of-the-industrial-robots/.
3. Karel Čapek, *R.U.R.* (Rossum's Universal Robots) (London/New York: Penguin Books, 2004). See also: "About R.U.R. (Rossum's Universal Robots)," Random House, accessed August 25, 2019, www.penguin randomhouse.com/books/286379/rur-rossums-universal-robots-by-karel -capek/9780141182087/.
4. "The Origin of the Word 'Robot,'" Science Friday, April 22, 2011, www .sciencefriday.com/segments/the-origin-of-the-word-robot/.
5. Voyen Koreis, "Čapek's R.U.R.," Booksplendour, accessed August 25, 2019, https://web.archive.org/web/20131223084729/http://www.booksplendour .com.au/capek/rur.htm.
6. Isaac Asimov, *I, Robot* (New York: Doubleday, 1950).
7. Isaac Asimov, "Runaround," *Astounding Science Fiction*, March 1942. See also: www.isfdb.org/cgi-bin/pl.cgi?57563.
8. Asimov, *I, Robot.*
9. Asimov first introduced the notion and need for a fourth law in the 1950s, but he didn't specify the law until 1985, in *Robots and Empire*. Isaac Asimov, *Robots and Empire* (New York: Doubleday, 1985).
10. "Atlas: The World's Most Dynamic Humanoid," Boston Dynamics, accessed August 25, 2019, www.bostondynamics.com/atlas.
11. "Rover," NASA Mars 2020 Mission, accessed August 25, 2019, https:// mars.nasa.gov/mars2020/mission/rover/.

CHAPTER 12

1. "'Leave It to Beaver' Series Debuts on CBS-TV 60 Years Ago This Hour #OnThisDay #OTC (Oct 4 1957)," RetroNewser Web Page, October 4, 2017, https://retronewser.com/2017/10/04/leave-it-to-beaver-series-debuts -on-cbs-tv-60-years-ago-this-hour-onthisday-otd-oct-4-1957/.

2. "Sputnik, 1957," Office of the Historian, Foreign Service Institute, US State Department, accessed August 24, 2019, https://history.state.gov /milestones/1953-1960/sputnik. See also: "The Launch of Sputnik, 1957," US Department of State Archive, accessed August 24, 2019, https://2001 -2009.state.gov/r/pa/ho/time/lw/103729.htm.

3. Although the American population was surprised by Sputnik, declassified information later revealed that the Eisenhower administration and the American military and intelligence services did anticipate that a Soviet satellite would likely achieve orbit no later than the end of 1957. See: Amy Ryan and Gary Keeley, "Sputnik and US Intelligence: The Warning Record," *Studies in Intelligence* 61, no. 3 (September 2017), https://www .cia.gov/library/center-for-the-study-of-intelligence/csi-publications/csi -studies/studies/vol-61-no-3/pdfs/sputnik-the-warning-record.pdf.

4. "Cold War: A Brief History, the Soviet Response," AtomicArchive.com, accessed August 24, 2019, www.atomicarchive.com/History/coldwar /page07.shtml.

5. Contrary to most Americans' understanding or recollection, there actually were a small number of attacks upon the American mainland during World War II, including Japan's shelling of the Ellwood oil field near Los Angeles, Japan's attack on Fort Stevens (an antiquated army base on the Oregon coastline), and Japan's launch of hydrogen balloon bombs, or "Fugos," into the jetstream, many of which made their way across the Pacific to various points in America. Evan Andrews, "5 Attacks on U.S. Soil During World War II," History, October 23, 2012, www.history.com/news/5-attacks-on -u-s-soil-during-world-war-ii.

6. "Sputnik (rocket)," WikiVisually, accessed August 24, 2019, https://wiki visually.com/wiki/Sputnik_(rocket).

7. CIA, "Khrushchev's 'We Will Bury You,'" written February 7, 1962. Declassified January 22, 2002, https://www.cia.gov/library/readingroom /docs/CIA-RDP73B00296R000200040087-1.pdf.

8. Jennifer Latson, "The Sad Story of Laika, the First Dog Launched into Space," *Time*, November 3, 2014, http://time.com/3546215/laika-1957/.

9. Constance McLaughlin Green and Milton Lomask, "Vanguard, a History," The NASA Historical Series, 1970, https://history.nasa.gov/SP-4202.pdf.

10. "Vanguard (Flopnik)," YouTube video, 1:14, posted by "superpowerplay," December 24, 2006, www.youtube.com/watch?time_continue=34&v=JK6 a6Hkp94o. See also: "Vanguard TV3 Failed Rocket Launch," YouTube

video, 0:53, posted by "NASA Langley CRGIS," August 11, 2009, www
.youtube.com/watch?v=zVeFkakURXM.

11. Alexandra Silver, "Kaputnik," *Time*, April 19, 2010, http://content
.time.com/time/specials/packages/article/0,28804,1982672_1982673
_1982707,00.html.

12. "Sputnik, 1957," Office of the Historian.

13. Sarah Loff, page editor, "Explorer 1 Overview," NASA, August 3, 2017,
www.nasa.gov/mission_pages/explorer/explorer-overview.html.

14. "The Birth of NASA: November 3, 1957–October 1, 1958," NASA History
Division, accessed August 24, 2019, https://history.nasa.gov/monograph
10/nasabrth.html.

15. "Where the Future Becomes Now," DARPA History and Timeline,
accessed August 24, 2019, https://www.darpa.mil/about-us/darpa-history
-and-timeline.

16. "The Birth of NASA: November 3, 1957–October 1, 1958," NASA History Division.

17. Audrey Watters, "How Sputnik Launched Ed-Tech: The National Defense
Education Act of 1958," Hack Education, June 20, 2015, http://hack
education.com/2015/06/20/sputnik. See also: Thomas C. Hunt, "National
Defense Education Act," *Encyclopedia Britannica*, accessed August 24, 2019,
www.britannica.com/topic/National-Defense-Education-Act.

18. For background, see: Nola Taylor Redd, "Yuri Gagarin: First Man in
Space," Space.com, October 12, 2018, www.space.com/16159-first-man
-in-space.html.

19. President John F. Kennedy, "Address at Rice University on the Nation's
Space Effort," Rice Stadium, September 12, 1962, John F. Kennedy Presidential Library and Museum, www.jfklibrary.org/learn/about-jfk/historic
-speeches/address-at-rice-university-on-the-nations-space-effort.

20. "The Belt and Road Initiative," World Bank, March 29, 2018, www.world
bank.org/en/topic/regional-integration/brief/belt-and-road-initiative.

21. "The Belt and Road Initiative," World Bank. See also: CEIC, Morgan Stanley Research, "Inside China's Plan to Create a Modern Silk Road," Morgan
Stanley, March 14, 2018, www.morganstanley.com/ideas/china-belt-and
-road.

22. Lucien O. Chauvin and Barbara Fraser, "South America Is Embracing
China's Science Silk Road," Nature.com, May 8, 2019, www.nature.com
/immersive/d41586-019-01127-4/index.html. See also: Nicholas Casey
and Clifford Krauss, "It Doesn't Matter If Ecuador Can Afford This

Dam, China Still Gets Paid," *Time*, December 24, 2018, www.nytimes
.com/2018/12/24/world/americas/ecuador-china-dam.html.

23. Valbona Zeneli, "Italy Signs On to Belt and Road Initiative: EU-China Relations at Crossroads?," The Diplomat, April 3, 2019, https://thediplomat
.com/2019/04/italy-signs-on-to-belt-and-road-initiative-eu-china-relations
-at-crossroads/.

24. Xu Wei, "China to Further Promote Entrepreneurship and Innovation," The
State Council of the People's Republic of China (English.Gov.CN), July 12,
2017, http://english.gov.cn/premier/news/2017/07/12/content_28147572308
6902.htm.

25. Kai-Fu Lee, *AI Superpowers* (Boston/New York: Houghton Mifflin Harcourt, 2018).

26. Xu Wei, "China to Further Promote Entrepreneurship and Innovation."

27. See: "Made in China 2025: Global Ambitions Built on Local Protections,"
US Chamber of Commerce, 2017, www.uschamber.com/sites/default/files
/final_made_in_china_2025_report_full.pdf.

28. "Made in China 2025," US Chamber of Commerce.

29. Matt Reynolds, "DeepMind's AI Beats World's Best Go Player in Latest Face-off," *New Scientist*, May 23, 2017, www.newscientist.com/article
/2132086-deepminds-ai-beats-worlds-best-go-player-in-latest-face-off/.

30. For a full translation, see: Graham Webster, Paul Triolo, Elsa Kania,
and Rogier Creemers, trans., "A Next Generation Artificial Intelligence
Development Plan," China Copyright and Media Web Page, July 20,
2017, https://chinacopyrightandmedia.wordpress.com/2017/07/20/a-next
-generation-artificial-intelligence-development-plan/.

31. Webster et al.

CHAPTER 13

1. "Timeline of Chinese History and Dynasties," Columbia University Asia
for Educators, accessed August 24, 2019, http://afe.easia.columbia.edu
/timelines/china_timeline.htm. See also: "The History of China: Dynasty/
Era Summary, Timeline," China Highlights, accessed August 24, 2019,
www.chinahighlights.com/travelguide/culture/china-history.htm.

2. See: Ilya Somin, "Remembering the Biggest Mass Murder in the History of
the World," *Washington Post*, August 3, 2016, www.washingtonpost.com
/news/volokh-conspiracy/wp/2016/08/03/giving-historys-greatest-mass
-murderer-his-due/?utm_term=.d3692bbd4344.

3. Somin, "Remembering the Biggest Mass Murder in the History of the World."

4. "China 'Social Credit System' Has Caused More Than Just Public Shaming (HBO)," YouTube video, 5:44, posted by VICE News, December 12, 2018, www.youtube.com/watch?v=Dkw15LkZ_Kw&t=125s.

5. Aris Teon, "Brainwashing the People—Mao Zedong, the Chinese Communist Party and the Politics of Thought Control," Greater China Journal, March 10, 2019, https://china-journal.org/2019/03/10/brainwashing-the -people-mao-zedong-the-chinese-communist-party-and-the-politics-of -thought-control/.

6. "Population of Cities in China (2019)," World Population Review, accessed August 24, 2019, http://worldpopulationreview.com/countries/china -population/cities/.

7. "US City Populations 2019," World Population Review, accessed August 24, 2019, http://worldpopulationreview.com/us-cities/.

8. "Population of Cities in China (2019)," World Population Review.

9. "US City Populations 2019," World Population Review.

10. Benjamin Haas, "More Than 100 Chinese Cities Now Above 1 Million People," The Guardian, accessed August 24, 2019, www.theguardian.com /cities/2017/mar/20/china-100-cities-populations-bigger-liverpool.

11. Haas, "More Than 100 Chinese Cities Now Above 1 Million People."

12. Paul Mozur, "Inside China's Dystopian Dreams: A.I., Shame and Lots of Cameras," New York Times, July 8, 2018, www.nytimes.com/2018/07/08 /business/china-surveillance-technology.html.

13. Epoch Newsroom, "Huawei and the Creation of China's Orwellian Surveillance State," Epoch Times, January 8, 2019, www.theepochtimes.com /huawei-and-the-creation-of-chinas-orwellian-surveillance-state_2747922 .html.

14. Mozur, "Inside China's Dystopian Dreams."

15. Will Kenton, "What Is Baidu?," Investopedia, May 7, 2019, www.investopedia .com/terms/b/baidu.asp.

16. Paige Leskin, "Here Are All the Major US Tech Companies Blocked Behind China's 'Great Firewall,'" Business Insider, March 23, 2019, www .businessinsider.com/major-us-tech-companies-blocked-from-operating -in-china-2019-5.

17. Staff, "Amazon's vs. Alibaba's Business Models: What's the Difference?," Investopedia, May 8, 2019, www.investopedia.com/articles/investing/061215 /difference-between-amazon-and-alibabas-business-models.asp.

18. Felix Todd, "What Is Tencent? Profiling China's Largest Social Media Conglomerate," NS Business, February 11, 2018, www.ns-businesshub .com/transport/what-is-tencent//.

19. Andrew Bloomenthal, "World's Top Ten Internet Companies," Investopedia, April 22, 2019, www.investopedia.com/articles/personal-finance/030415 /worlds-top-10-internet-companies.asp.

20. Karen Hao, "Three Charts Show How China's AI Industry Is Propped Up by Three Companies," *MIT Technology Review*, January 22, 2019, www .technologyreview.com/s/612813/the-future-of-chinas-ai-industry-is-in -the-hands-of-just-three-companies/.

21. Arjun Kharpal, "Huawei Says It Would Never Hand Data to China's Government. Experts Say It Wouldn't Have a Choice," CNBC, March 4, 2019, www.cnbc.com/2019/03/05/huawei-would-have-to-give-data-to-china -government-if-asked-experts.html.

22. Edwin Chan, David Ramli, and Lulu Yilum, "The Great Firewall of China," Bloomberg News, November 5, 2018, www.bloomberg.com /quicktake/great-firewall-of-china.

23. Sophie Perryer, "China's Social Credit System Awards Points to Citizens Who Conform," The New Economy, May 22, 2019, www.theneweconomy.com /strategy/116498. See also: Lily Kuo, "China Bans 23m from Buying Travel Tickets as Part of 'Social Credit' System," *The Guardian*, March 1, 2019, www.theguardian.com/world/2019/mar/01/china-bans-23m-discredited -citizens-from-buying-travel-tickets-social-credit-system.

24. Kai-Fu Lee, *AI Superpowers* (Boston/New York: Houghton Mifflin Harcourt, 2018).

25. Lee, *AI Superpowers*.

26. Lee, *AI Superpowers*.

27. Chris Udemans, "Blacklists and Redlists: How China's Social Credit System Actually Works," TechNode, October 23, 2018, https://technode .com/2018/10/23/china-social-credit/. See also: Rosie Perper, "Chinese Dog Owners Are Being Assigned a Social Credit Score to Keep Them in Check—and It Seems to Be Working," *Business Insider*, October 26, 2018, www.businessinsider.com/china-dog-owners-social-credit-score-2018-10.

28. Udemans, "Blacklists and Redlists."

29. Kenneth Pletcher, "One-Child Policy, Chinese Government Program," *Encyclopedia Britannica*, accessed September 7, 2019, www.britannica.com /topic/one-child-policy.

30. Zheping Huang, "China's Zheping Most Popular App Is a Propaganda Tool Teaching Xi Jinping Thought," *South China Morning Post*, February 14, 2019, www.scmp.com/tech/apps-social/article/2186037/chinas-most-popular-app-propaganda-tool-teaching-xi-jinping-thought. See also: Rita Liao, "A Government Propaganda App Is Going Viral in China," TechCrunch, February 2019, https://techcrunch.com/2019/02/01/china-propaganda-app/.

31. Victor C. Falkenheim and Chiao-Min Hsieh, "Xinjiang," *Encyclopedia Britannica*, accessed August 25, 2019, www.britannica.com/place/Xinjiang.

32. Xinjiang borders the Chinese autonomous region of Tibet and eight foreign countries: the Altai Republic of Russia, Kazakhstan, Kyrgyzstan, Tajikistan, Afghanistan, Pakistan, India, Mongolia.

33. Tara Francis Chan, "As the U.S. Targets China's 'Concentration Camps,' Xinjiang's Human Rights Crisis Is Only Getting Worse," *Newsweek*, May 22, 2019, www.newsweek.com/xinjiang-uyghur-crisis-muslim-china-1398782.

34. Austin Ramzy and Chris Buckley, "The Xinjiang Papers 'Absolutely No Mercy': Leaked Files Expose How China Organized Mass Detentions of Muslims," *New York Times*, November 16, 2019, https://www.nytimes.com/interactive/2019/11/16/world/asia/china-xinjiang-documents.html.

35. Eva Xiao and Pak Yiu, "Uighur Tombs, Cemeteries Cleared in Xinjiang," *Asia Times*, October 12, 2019. https://asiatimes.com/2019/10/uighur-tombs-cemeteries-cleared-in-xinjiang/.

36. Paul Mozur, "One Month, 500,000 Face Scans: How China Is Using A.I. to Profile a Minority," *New York Times*, April 14, 2019, www.nytimes.com/2019/04/14/technology/china-surveillance-artificial-intelligence-racial-profiling.html. See also: Eric Lutz, "China Has Created a Racist A.I. to Track Muslims," *Vanity Fair*, April 15, 2019, www.vanityfair.com/news/2019/04/china-created-a-racist-artificial-intelligence-to-track-muslims.

37. Mozur, "One Month"; Lutz, "China Has Created a Racist A.I."

38. Paul Mozur, Jonah M. Kessel, and Melissa Chan, "Made in China, Exported to the World: The Surveillance State," *New York Times*, April 24, 2019, www.nytimes.com/2019/04/24/technology/ecuador-surveillance-cameras-police-government.html. See also: Charles Rollet, "Ecuador's All-Seeing Eye Is Made in China," *Foreign Policy*, August 9, 2018, https://foreignpolicy.com/2018/08/09/ecuadors-all-seeing-eye-is-made-in-china/.

39. Mozur et al., "Made in China, Exported to the World: The Surveillance State."

40. Mozur et al., "Made in China, Exported to the World: The Surveillance State."

41. Raymond Zhong, "Who Owns Huawei? The Company Tried to Explain. It Got Complicated," *New York Times*, April 25, 2019, www.nytimes .com/2019/04/25/technology/who-owns-huawei.html. See also: Christopher Balding and Donald C. Clarke, "Who Owns Huawei?," SSRN, April 17, 2019, https://papers.ssrn.com/sol3/papers.cfm?abstract_id=3372669.

42. John Edwards, "5G Versus 4G: How Speed, Latency and Application Support Differ," NetworkWorld, January 7, 2019, www.networkworld .com/article/3330603/5g-versus-4g-how-speed-latency-and-application -support-differ.html.

43. Katharina Buchholz, "Which Countries Have Banned Huawei?," Statista, August 19, 2019, www.statista.com/chart/17528/countries-which-have -banned-huawei-products/.

44. Stu Woo and Kate O'Keeffe, "Washington Asks Allies to Drop Huawei," *Wall Street Journal*, November 23, 2019, www.wsj.com/articles /washington-asks-allies-to-drop-huawei-1542965105.

45. Gregory C. Allen, "Understanding China's AI Strategy," Center for a New American Security (CNAS), February 6, 2019, www.cnas.org/publications /reports/understanding-chinas-ai-strategy.

46. People's Republic of China, "Position Paper of China," submitted to the United Nations Group of Governmental Experts of the High Contracting Parties to the Convention on Prohibitions or Restrictions on the Use of Certain Conventional Weapons Which May Be Deemed to Be Excessively Injurious or to Have Indiscriminate Effects, April 11, 2018, www.unog .ch/80256EDD006B8954/(httpAssets)/E42AE83BDB3525D0C125826 C0040B262/$file/CCW_GGE.1_2018_WP.7.pdf.

47. Richard A. Bitzinger, "How China Weaponized Overseas Arms Sales," *Asia Times*, April 16, 2019, www.asiatimes.com/2019/04/opinion/how-china -weaponizes-overseas-arms-sales/.

CHAPTER 14

1. The 15 independent nations that formed after the fall of the Soviet Union are: Armenia, Azerbaijan, Belarus, Estonia, Georgia, Kazakhstan, Kyrgyzstan, Latvia, Lithuania, Moldova, Russia, Tajikistan, Turkmenistan, Ukraine, Uzbekistan. See: Justin Burke, "Post-Soviet World: What You Need to Know About the 15 States," *The Guardian*, accessed August 25,

2019, www.theguardian.com/world/2014/jun/09/-sp-profiles-post-soviet -states. See also: "Post-Soviet States," Wikipedia, accessed August 25, 2019, https://en.wikipedia.org/wiki/Post-Soviet_states.

2. "The Largest Countries in the World," World Atlas, accessed August 6, 2019, www.worldatlas.com/articles/the-largest-countries-in-the-world-the -biggest-nations-as-determined-by-total-land-area.html.

3. Yehuda Z Blum, "Russia Takes Over the Soviet Union's Seat at the United Nations," *Kaleidoscope/European Journal of International Law*, accessed May 28, 2019, www.ejil.org/article.php?article=2045&issue=102.

4. "Economy After the Break Up of the Soviet Union," Facts and Details, accessed August 25, 2019, http://factsanddetails.com/russia/Economics _Business_Agriculture/sub9_7b/entry-5167.html.

5. Karen Dawisha, *Putin's Kleptocracy* (New York: Simon & Schuster, 2014).

6. Rob Wile, "Is Putin Secretly the World's Richest Man?," Money, January 23, 2017, http://money.com/money/4641093/vladimir-putin-net-worth/.

7. Simon Saradzhyan, "5 Years Since Russia's Intervention in Ukraine: Has Putin's Gamble Paid Off?," Harvard Kennedy School/Russia Matters, March 14, 2019, www.russiamatters.org/analysis/5-years-russias -intervention-ukraine-has-putins-gamble-paid.

8. James Vincent, "Putin Says the Nation That Leads in AI 'Will be the Ruler of the World,'" The Verge, September 4, 2017, www.theverge.com/2017 /9/4/16251226/russia-ai-putin-rule-the-world. See also: Staff, "Whoever Leads in AI Will Rule the World: Putin to Russian Children on Knowledge Day," RT Question More, September 1, 2017, www.rt.com/news/401731 -ai-rule-world-putin/.

9. "Decree of the President of the Russian Federation," trans. and ed. Margarity Konaev, Alexandra Vreeman, and Ben Murphy, Center for Security and Emerging Technology, October 28, 2019, https://cset.george town.edu/wp-content/uploads/t0060_Russia_AI_strategy_EN-1.pdf. See also: Margarita Konaev, "Thoughts on Russia's AI Strategy," Georgetown University Walsh School of Foreign Service, October 30, 2019, https://cset .georgetown.edu/2019/10/30/russias-ai-strategy/.

10. Samuel Bendett, "In AI, Russia Is Hustling to Catch Up," Defense One, April 4, 2018, www.defenseone.com/ideas/2018/04/russia-races-forward -ai-development/147178/.

11. Samuel Bendett, "Russia Is Poised to Surprise the US in Battlefield Robot-ics," Defense One, January 25, 2018, www.defenseone.com/ideas/2018/01 /russia-poised-surprise-us-battlefield-robotics/145439/.

12. Mark Bennetts, "Putin: Syria War Is Priceless for Testing Our New Weapons," *Times* (London), June 8, 2018, www.thetimes.co.uk/article/putin-syria-war-is-priceless-for-testing-our-new-weapons-qkz3qsdqw.

13. Matt Bartlett, "The AI Arms Race in 2019," Towards Data Science, January 28, 2019, https://towardsdatascience.com/the-ai-arms-race-in-2019-fdca07a086a7.

14. Hayley Evans and Natalie Salmanowitz, "Lethal Autonomous Weapons Systems: Recent Developments," *Lawfare*, March 7, 2019, www.lawfareblog.com/lethal-autonomous-weapons-systems-recent-developments.

15. "Russia," Reporters Without Borders, accessed August 18, 2019, www.rsf.org/en/russia.

16. Charlotte Jee, "Russia Wants to Cut Itself Off from the Global Internet. Here's What That Really Means," *MIT Technology Review*, March 21, 2019, www.technologyreview.com/s/613138/russia-wants-to-cut-itself-off-from-the-global-internet-heres-what-that-really-means/.

17. James Andrew Lewis, "Reference Note on Russian Communications Surveillance," Center for Strategic & International Studies (CSIS), April 18, 2014, www.csis.org/analysis/reference-note-russian-communications-surveillance.

18. Willian McKinney, "Meet SORM, Russia's Anti-Hacking and Surveillance System," Edgy, January 12, 2017, https://edgy.app/russia-sorm-election.

19. "Russia to Arm Police with AR Face Recognition Glasses by 2020," *Moscow Times*, May 24, 2019, www.themoscowtimes.com/2019/05/24/russia-to-arm-police-with-ar-face-recognition-glasses-by-2020-a65720. See also: "Russia Tests Facial Recognition Cameras in Moscow Ahead of World Cup," *Moscow Times*, April 18, 2018, www.themoscowtimes.com/2018/04/18/russia-tests-facial-recognition-cameras-moscow-ahead-world-cup-a61201.

20. Sophia Porotsky, "Analyzing Russian Information Warfare and Influence Operations," Global Security Review, February 8, 2018, https://globalsecurityreview.com/cold-war-2-0-russian-information-warfare/.

21. Andrew Kramer, "Russian General Pitches 'Information' Operations as a Form of War," *New York Times*, March 2, 2019, www.nytimes.com/2019/03/02/world/europe/russia-hybrid-war-gerasimov.html.

22. "The Military Doctrine of the Russian Federation" (press release translated from Russian), the Embassy of the Russian Federation to the United Kingdom of Great Britain and Northern Ireland, June 29, 2015, https://rusemb.org.uk/press/2029.

23. Hannah Arendt, "Hannah Arendt: From an Interview," *New York Review of Books*, October 26, 1978, www.nybooks.com/articles/1978/10/26/hannah -arendt-from-an-interview/.

24. See: Matt Apuzzo and Adam Satariano, "Russia Is Targeting Europe's Elections. So Are Far-Right Copycats," *New York Times*, May 12, 2019, www .nytimes.com/2019/05/12/world/europe/russian-propaganda-influence -campaign-european-elections-far-right.html.

25. See: "U.S. Senate Select Committee on Intelligence Report, July 3, 2018," US Senate Publications, accessed August 25, 2019, https://www.intelligence .senate.gov/publications/committee-findings-2017-intelligence-community -assessment. See also: "The St. Petersburg Troll Factory Targets Elections from Germany to the United States," EU vs Disinfo, April 2, 2019, https:// euvsdisinfo.eu/the-st-petersburg-troll-factory-targets-elections-from -germany-to-the-united-states/.

26. "U.S. Senate Select Committee on Intelligence Report"; "The St. Petersburg Troll Factory Targets Elections."

27. Zachary Basu, "Read the Full Transcript of Mueller's Statement," Axios, May 29, 2019, www.axios.com/mueller-statement-transcript-c1609594 -ff8d-4824-abec-4f7415833129.html.

CHAPTER 15

1. Mark Cartwright, "Cleisthenes," Ancient History Encyclopedia, April 8, 2016, www.ancient.eu/Cleisthenes/. See also: Mark Cartwright, "Athenian Democracy," Ancient History Encyclopedia, April 3, 2018, www.ancient .eu/Athenian_Democracy/.

2. The Macedonian conquest of Greece occurred under King Philip II, who ruled Macedonia from 359 to 336 BCE. Although aspects of Athenian democracy survived the city's conquest and occupation, it's arguable that all aspects of the Athenian model of government had been abandoned or abolished by the time of Alexander the Great's death in 323 BCE. See: "Philip of Macedon — King of Macedonia and Conqueror of Illyria, Thrace, and Greece," Ancient Macedonia, accessed August 19, 2019, www.ancient macedonia.com/PhilipofMacedon.html.

3. "The Petition of Right, 1628," Constitution Society, accessed August 19, 2019, www.constitution.org/eng/petright.htm.

4. "Bill of Rights, 1689," Constitution Society, accessed August 19, 2019, www.constitution.org/eng/eng_bor.htm.

5. "Thirteen Colonies Population (1710–1770)," World Population Review, accessed May 27, 2019, http://worldpopulationreview.com/states/thirteen -colonies/.

6. "A Summary of the 1765 Stamp Act," Colonial Williamsburg, accessed May 27, 2019, www.history.org/history/teaching/tchcrsta.cfm. See also: "A Report on Reaction to the Stamp Act, 1765," The Gilder Lehrman Institute of American History, accessed May 27, 2019, www.gilderlehrman .org/content/report-reaction-stamp-act-1765.

7. "A Summary of the 1765 Stamp Act"; "A Report on Reaction to the Stamp Act, 1765."

8. The Declaration of Independence, 1776, Office of the Historian, accessed May 27, 2019, https://history.state.gov/milestones/1776-1783/declaration.

9. "Declaration of Independence: A Transcription," National Archives, accessed May 27, 2019, www.archives.gov/founding-docs/declaration -transcript.

10. A surprisingly small number of casualties actually occurred on the Revolutionary battlefields. But many thousands more died from their battle wounds, widespread disease, poor medical care, and sparse food and fresh water. Although accurate death records are difficult to compile.

11. "The Treaty of Paris," ConstitutionFacts.com, accessed August 25, 2019, www.constitutionfacts.com/us-declaration-of-independence/treaty-of -paris/.

12. "The Constitution of the United States: A Transcription," National Archives, accessed August 25, 2019, www.archives.gov/founding-docs /constitution-transcript.

13. "The Bill of Rights: A Transcription," National Archives, accessed August 25, 2019, www.archives.gov/founding-docs/bill-of-rights-transcript.

14. Max Rosner, "Democracy," Our World in Data, accessed August 18, 2019, https://ourworldindata.org/democracy.

15. This is frequently cited as a quote by Churchill, but it's questionable where and when it was actually stated, if at all. Nonetheless, it's consistent, as Churchill often commented on the shortcomings of democracy. See: Michael Richards, "History Detectives—Red Herrings: Famous Words Churchill Never Said," International Churchill Society, accessed August 19, 2019, https://winstonchurchill.org/publications/finest-hour /finest-hour-141/history-detectives-red-herrings-famous-words -churchill-never-said/.

16. Tim Sharp, "Right to Privacy: Constitutional Rights and Privacy Laws," Live Science, June 12, 2013, www.livescience.com/37398-right-to-privacy .html.

17. "The National Artificial Intelligence Research and Development Strategic Plan," National Science and Technology Council, October 2017, www .nitrd.gov/PUBS/national_ai_rd_strategic_plan.pdf.

18. See: Future of Life Institute, https://futureoflife.org/.

19. See: Future of Humanity Institute, www.fhi.ox.ac.uk/.

20. See: Machine Intelligence Research Institute, https://intelligence.org/.

21. "The National Artificial Intelligence Research and Development Strategic Plan."

22. "Summary of the 2018 White House Summit on Artificial Intelligence for American Industry," The White House Office of Science and Technology Policy, May 10, 2018, www.whitehouse.gov/wp-content /uploads/2018/05/Summary-Report-of-White-House-AI-Summit.pdf. And: "Artificial Intelligence for the American People," WhiteHouse .gov, May 10, 2018, www.whitehouse.gov/briefings-statements/artificial -intelligence-american-people/.

23. "Summary of the 2018 White House Summit on Artificial Intelligence for American Industry"; "Artificial Intelligence for the American People."

24. 115th Congress (2017–2018), "National Security Commission on Artificial Intelligence Act of 2018," Congress.gov, www.congress.gov/bill /115th-congress/senate-bill/2806/all-info.

25. Justin Doubleday, "Top Tech Execs Named to New National Security Commission on Artificial Intelligence," Inside Defense, January 10, 2019, https://insidedefense.com/insider/top-tech-execs-named-new-national -security-commission-artificial-intelligence.

26. "Executive Order on Maintaining American Leadership in Artificial Intelligence," WhiteHouse.gov, February 11, 2019, www.whitehouse.gov /presidential-actions/executive-order-maintaining-american-leadership -artificial-intelligence/.

27. "Executive Order on Maintaining American Leadership in Artificial Intelligence."

28. "About the Department of Defense (DoD)/Mission," Department of Defense Archive, accessed May 27, 2019, https://archive.defense.gov /about/#mission.

29. "DoD Personnel, Workforce Reports & Publications," DMDC, accessed May 27, 2019, www.dmdc.osd.mil/appj/dwp/dwp_reports.jsp.

30. "Summary of the 2018 Department of Defense Artificial Intelligence Strategy," DoD, 2018, https://media.defense.gov/2019/Feb/12/2002088963/-1/-1/1/summary-of-dod-ai-strategy.pdf.

31. Mark Landler and Kate Kelly, "Davos in the Desert. A Saudi Prince's Glittering Showcase Is Stained by a Grisly Accusation," *New York Times*, October 13, 2018, www.nytimes.com/2018/10/13/world/middleeast/saudi-arabia-conference-crown-prince-mohammed.html.

32. Anand Giridharadas, "Silicon Valley's Saudi Arabia Problem," *New York Times*, October 15, 2018, www.nytimes.com/2018/10/12/opinion/silicon-valley-saudi-arabia.html.

33. Theodore Schleifer, "Silicon Valley Is Awash in Chinese and Saudi Cash—and No One Is Paying Attention (Except Trump)," Vox, May 1, 2019, www.vox.com/recode/2019/5/1/18511540/silicon-valley-foreign-money-china-saudi-arabia-cfius-firrma-geopolitics-venture-capital.

34. Schleifer, "Silicon Valley Is Awash in Chinese and Saudi Cash."

35. Taylor Hatmaker, "Saudi Arabia Bestows Citizenship on a Robot Named Sophia," TechCrunch, October 26, 2017, https://techcrunch.com/2017/10/26/saudi-arabia-robot-citizen-sophia/.

36. See: Hanson Robotics, accessed May 28, 2019, www.hansonrobotics.com/sophia/.

37. See: Brigit Katz, "Why Saudi Arabia Giving a Robot Citizenship Is Firing People Up," *Smithsonian*, November 2, 2017, www.smithsonianmag.com/smart-news/saudi-arabia-gives-robot-citizenshipand-more-freedoms-human-women-180967007/.

38. Katz, "Why Saudi Arabia Giving a Robot Citizenship Is Firing People Up."

39. Hillary Leung, "What to Know About Absher, Saudi Arabia's Controversial 'Woman-Tracking' App," *Time*, February 19, 2019, http://time.com/5532221/absher-saudi-arabia-what-to-know/. See also: Editorial Board, "A Tool of Oppression: Apple, Google Should Take Action Against Saudi App," *Pittsburgh Post-Gazette*, May 13, 2019, www.post-gazette.com/opinion/editorials/2019/05/13/Saudi-Arabia-Absher-app-oppression-womens-rights-Apple-Google/stories/201905130019.

40. "The History of the European Union," European Union, accessed May 28, 2019, https://europa.eu/european-union/about-eu/history_en.

41. "The History of the European Union."

42. "Presidency Conclusions, Copenhagen European Council, June 21–22, 1993," www.europarl.europa.eu/enlargement/ec/pdf/cop_en.pdf.

43. "EU Citizenship," European Union, accessed May 28, 2019, https://europa .eu/european-union/about-eu/eu-citizenship_en.

44. The European Union's Science and Technology Services, "Declaration on Cooperation on Artificial Intelligence," European Commission, April 2018, https://ec.europa.eu/jrc/communities/en/node/1286/document/eu -declaration-cooperation-artificial-intelligence.

45. Bryce Goodman and Seth Flaxman, "European Union Regulations on Algorithmic Decision-Making and a 'Right to Explanation,'" accessed May 28, 2019, https://arxiv.org/pdf/1606.08813v3.pdf.

46. "Alleged Global Google Privacy Leak: 'GDPR Workaround' Could Incur $5.4B Fine," *Forbes*, April 20, 2020, https://www.forbes.com /sites/johnkoetsier/2019/09/04/alleged-global-google-privacy-leak-gdpr -workaround-could-incur-27b-fine/#5416caf32a02.

47. "Communication Artificial Intelligence for Europe," European Commission, April 25, 2018, https://ec.europa.eu/digital-single-market/en /news/communication-artificial-intelligence-europe.

48. "Communication Artificial Intelligence for Europe."

49. European Commission High-Level Expert Group on Artificial Intelligence Set Up by the European Commission, "Ethics Guidelines for Trustworthy AI," European Commission, April 8, 2019, https://ec.europa.eu /digital-single-market/en/news/ethics-guidelines-trustworthy-ai.

50. House of Lords Select Committee on Artificial Intelligence, "AI in the UK: Ready, Willing and Able?," Parliament Publications, April 16, 2018, https://publications.parliament.uk/pa/ld201719/ldselect/ldai/100/100.pdf.

51. See: The Alan Turing Institute, www.turing.ac.uk/.

52. See: Future of Humanity Institute, www.fhi.ox.ac.uk/.

53. House of Lords Select Committee on Artificial Intelligence, "AI in the UK: Ready, Willing and Able?"

54. "Theresa May's 2018 Davos Address in Full," World Economic Forum, January 25, 2018, www.weforum.org/agenda/2018/01/theresa-may-davos -address/.

55. "Artificial Intelligence: 'Making France a Leader,'" Gouvernement.fr, March 30, 2019, www.gouvernement.fr/en/artificial-intelligence-making-france -a-leader. See also: Mathieu Rosemain and Michel Rose, "France to Spend $1.8 Billion on AI to Compete with U.S., China," Reuters, March 29, 2018, www.reuters.com/article/us-france-tech/france-to-spend-1-8-billion-on-ai -to-compete-with-u-s-china-idUSKBN1H51XP.

56. CIFAR, "CIFAR Launches AI & Society Workshop Call for Proposals," CIFAR, accessed May 28, 2019, www.cifar.ca/cifarnews/2018/04/19/cifar -launches-ai-society-workshop-call-for-proposals.

57. See: "Canada First to Adopt Strategy for Artificial Intelligence," UNESCO Media Services, March 2017, www.unesco.org/new/en/media-services/single -view/news/canada_first_to_adopt_strategy_for_artificial_intelligence/. See also: CIFAR, "CIFAR Pan-Canadian Artificial Intelligence Strategy," CIFAR, accessed May 28, 2019, www.cifar.ca/ai/pan-canadian-artificial -intelligence-strategy.

58. CIFAR, "CIFAR Pan-Canadian Artificial Intelligence Strategy."

59. The Quebec G7 Summit was the sixth time since 1981 that Canada hosted the meetings. It included leaders from Canada, France, Germany, Italy, Japan, the United Kingdom, and the United States. See: "G7 Summit Ends in Disarray as Trump Abandons Joint Statement," BBC News, June 10, 2018, www.bbc.com/news/world-us-canada-44427660.

60. See: "G7 Multistakeholder Conference on Artificial Intelligence," Government of Canada, accessed May 28, 2019, www.ic.gc.ca/eic/site/133.nsf /eng/home.

61. George Nott, "Budget 2018: Funding Boost for AI and Machine Learning Projects," CIO, May 8, 2018, www.cio.com.au/article/640928/budget -2018-funding-boost-ai-machine-learning-projects/.

62. Edward Pollitt, "Budget 2018: National AI Ethics Framework on the Way," *Information Age*, May 10, 2018, https://ia.acs.org.au/article/2018 /budget-2018--ai-boost-with-an-ethical-focus.html.

63. "Huawei and ZTE Handed 5G Network Ban in Australia," BBC News, August 23, 2018, www.bbc.com/news/technology-45281495.

64. Saheli Roy Choudhury, "Former Australian PM Turnbull Explains Why His Government Banned Huawei, ZTE from Selling 5G Equipment," CNBC, March 28, 2019, www.cnbc.com/2019/03/28/malcolm-turnbull -on-australias-decision-to-ban-chinas-huawei-and-zte.html.

65. Matthew Brockett, "New Zealand Bans China's Huawei from 5G Wireless Networks," Bloomberg News, November 27, 2018, www.bloomberg .com/news/articles/2018-11-28/new-zealand-bans-china-s-huawei-from -5g-wireless-networks.

66. "National and International AI Strategies," Future of Life Institute, accessed August 25, 2019, https://futureoflife.org/national-international -ai-strategies/?cn-reloaded=1.

CHAPTER 16

1. OpenAI is a nonprofit company founded in 2015 by Elon Musk and other investors with an expressed mission to ensure that artificial intelligence remains safe and that it benefits all of humanity. OpenAI, accessed August 25, 2019, https://openai.com/.
2. OpenAI, https://openai.com/.
3. "Better Language Models and Their Implications," OpenAI, accessed August 25, 2019, https://openai.com/blog/better-language-models/. And see: "GPT-2: 6 Month Follow-up," OpenAI, accessed August 25, 2019, https://openai.com/blog/gpt-2-6-month-follow-up/.

INDEX